ADDITI[illegible] PRAISE FOR

# THE [illegible]



"This delightful, well-writter[illegible] bet-ter time, given the high glo[illegible] ergy production and the need to store it."

—[illegible] *Today*

"Henry Schlesinger's fascinating and superbly researched history of the battery is the story of civilization as we know it. *The Battery* illuminates in compellingly rich detail the scientists and entrepreneurs responsible for so much of the technology we take for granted today."

—Michael Belfiore, author of *The Department of Mad Scientists*

"Author Henry Schlesinger is playful and intelligent and obscenely well read. Just as a cracker-sized battery powers a cell phone for days, so does Schlesinger's wit enliven an unlikely topic—the history of the ever-shrinking, ever-more-potent battery."

—Richard Zacks, author of *An Underground Education* and the bestseller *The Pirate Hunter*

"The battery touches all our lives. The development of portable power mirrors our own society's development, and where we will be in the future will largely be a function of how portable and long-lasting our power sources will be. If the clues to our future are contained in the past, then the story of the battery's development points toward the technological frontiers ahead."

—Frank Vizard, technology journalist

"Under the guise of reintroducing us to the humble battery, Henry Schlesinger has actually written a fascinating history of human innovation spanning three millennia. *The Battery* is as sparky and as versatile as its subject."

—Ken Jennings, *New York Times* bestselling author of *Brainiac* and the greatest champion in *Jeopardy!* history

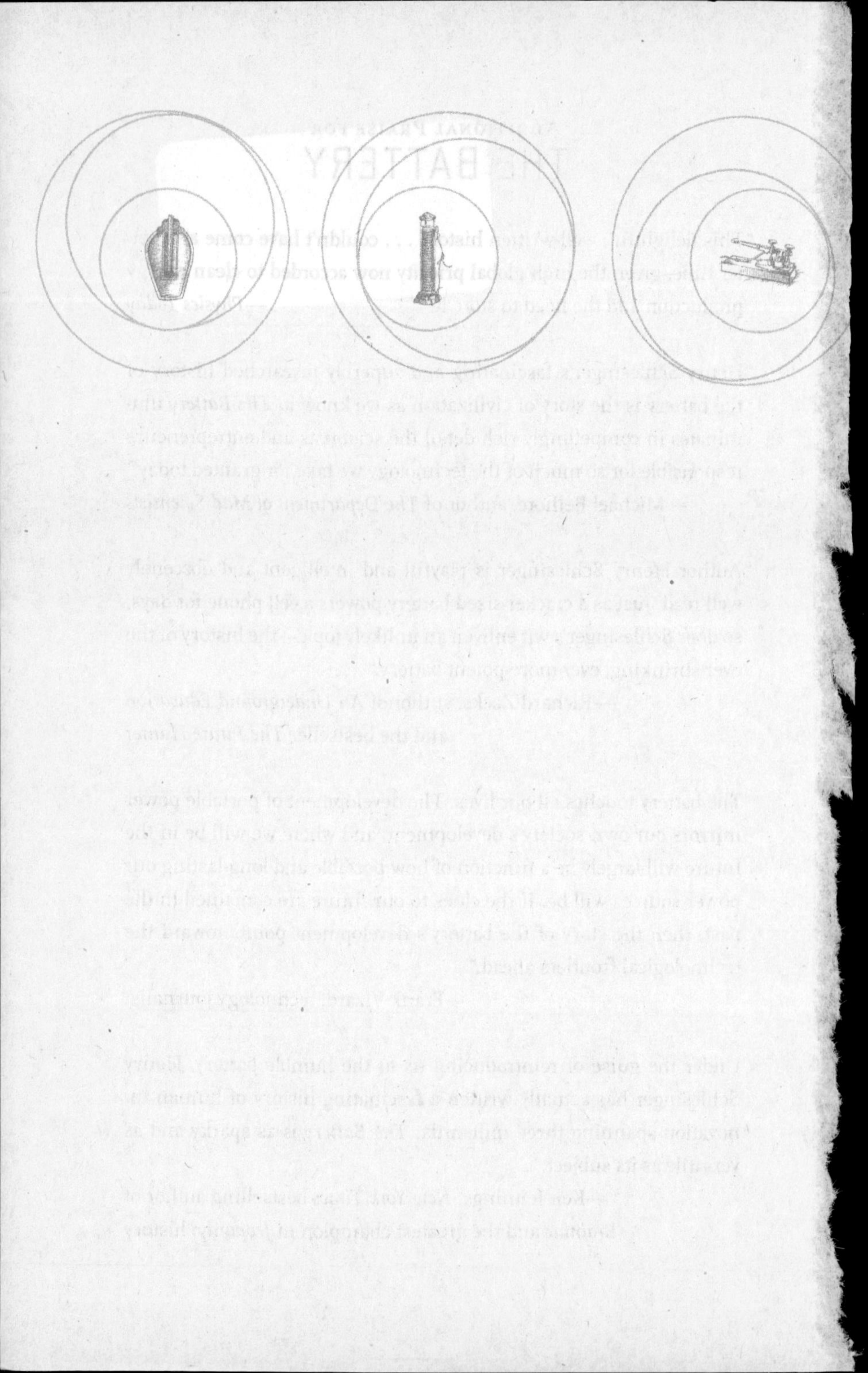

ADDITIONAL PRAISE FOR

# THE BATTERY

"This delightful and well-written history . . . couldn't have come at a better time, given the high global priority now accorded to clean energy production and the need to store it." —*Physics Today*

"Henry Schlesinger's fascinating and superbly researched history of the battery is the story of civilization as we know it. *The Battery* illuminates in compelling, rich detail the scientists and entrepreneurs responsible for so much of the technology we take for granted today."
—Michael Belfiore, author of *The Department of Mad Scientists*

"Author Henry Schlesinger is playful and intelligent and deceptively well read. Just as a cigarette-sized battery powers a cell phone for days, so does Schlesinger's wit enliven an unlikely topic—the history of this ever-shrinking, ever more potent source."
—Richard Zacks, author of *An Underground Education* and the bestseller *The Pirate Hunter*

"The battery touches all of our lives. The development of portable power mirrors our own society's development and where we will be in the future will largely be a function of how portable and long-lasting our power source will be. If the clues to our future are contained in the past, then the story of the battery's development points toward the technological frontiers ahead."
—Harry [illegible], technology journalist

"Under the guise of reintroducing us to the humble battery, Henry Schlesinger has actually written a fascinating history of human innovation spanning three millennia. *The Battery* is as sparky and as versatile as its subject."
—Ken Jennings, *New York Times* bestselling author of *Brainiac* and the greatest champion in *Jeopardy!* history

# THE BATTERY

## *How Portable Power Sparked a Technological Revolution*

Henry Schlesinger

HARPER

NEW YORK • LONDON • TORONTO • SYDNEY

*To Elisabeth Dyssegaard, Kathryn Whitenight, Jud Laghi, Charlie Scuilla, and Melissa Suzanne, for their patience.*

*I'd also like to thank all of the people who helped me along the way. They're a diverse group, and I'm certain that I've inadvertently missed some of them. However, I am grateful to the assistance given to me by Susan Hendrix, Larry Steckler, Tracey Thiele, Dennis Klein, Janell Mirochna, Mindy Rosewitz, Wendy Rejan, Corey Crell, and Donna Frazier Schmitt. And, of course, to Chris Costello, for his most excellent illustrations.*

HARPER

A hardcover edition of this book was published in 2010 by HarperCollins Publishers.



FIRST HARPER PAPERBACK PUBLISHED 2011.



*Designed by Mary Austin Speaker*

The Library of Congress has catalogued the hardcover edition as follows:
Schlesinger, Henry R.
The battery : how portable power sparked a technological revolution / Henry Schlesinger. —1st ed.
p. cm.
Includes bibliographical references and index.
ISBN 978-0-06-144293-3 (alk. paper)
1. Electric batteries—Design and construction—History. 2. Storage batteries—Design and construction—History. I. Title.
TK2901.S27 2010
621.31'242—dc22 2009034303

ISBN 978-0-06-144294-0 (pbk.)

23 24 25 26 27 LBC 8 7 6 5 4

*Faith is a fine invention*
*For gentlemen who see;*
*But microscopes are prudent*
*In an emergency!*

—Emily Dickinson

# CONTENTS

# INTRODUCTION

## *History in Real Time*

> *"Nothing is too wonderful to be true if it be consistent with the laws of nature."*
>
> —*Michael Faraday*

This book has something of an unlikely origin. It was during the course of working on a book about espionage that I became interested in batteries. Because changing the batteries of a piece of spy gear in the field is often inconvenient, if not altogether impossible, power sources are considered a critical component when it comes to intelligence gathering. The engineers and scientists who dream up all the fancy spy gear spend a lot of time worrying about batteries. While surprising, it also made perfect sense. James Bond was never seen popping into a drugstore for a couple of AAs to power up his gadgets, but *something* had to power them.

The espionage book was a lengthy project, but during my downtime I began making notes about batteries on small index cards. One slim stack of cards very quickly turned into two, then grew into four, and soon expanded into eight. The answer to each question seemed to prompt four more questions. Clearly there was more to batteries than we generally realize.

A little more research revealed that there was almost nothing written about batteries for the nontechnical reader. Of course, it's possible to find individual books on the chemistry, physics, history, and electronics of batteries, but these are overwhelmingly intended for technical or scientific professionals and academics. They are almost always very narrowly focused and, to be perfectly blunt, pretty dry stuff. Conversely, the vast majority of breathless prose churned out about consumer gadgets touches only briefly on the topic of batteries. All the action is in the user interface—the display, the keypad, the speed, and the apps. By comparison, batteries are generally regarded as somewhat dreary necessities. Even the most diehard tech geeks I know have a hard time mustering enthusiasm for battery technology.

THE FACT IS, BATTERIES NOT only power our current technologically advanced and portable age, but are also largely responsible for virtually all of the early basic scientific research that made today's gadgets and gizmos possible. Batteries quite literally powered much of the basic science that led to the consumer technology they power today. Without batteries, not only would our cell phones and other gadgets not work; in all likelihood the technology on which they are based would not exist. This is the kind of elegant, circular dynamic that is irresistible to a writer.

However, there is another important aspect. Since batteries are an enabling technology, it's impossible to understand their significance without providing scientific, historical, and technological context. In writing this book there was very much a sense of being let loose in history's candy store. Pick a subject, from home appliances to the world's battlefields, and you'll find batteries powering up an increasingly sophisticated technology.

And there were more surprises. From the very beginning, literally within weeks of publication describing the first "modern" battery in 1800, scientists began making improvements on the initial design. The quiet, steady evolution to increase battery power and extend life started before they even understood exactly how they worked or the true nature of electricity.

The intent of this book is to draw together those disparate and seemingly unrelated elements to tell the story. If there are detours, it is only because the facts uncovered were either too interesting or too much fun to leave out. As an author, I'd like to believe this is the first book in which Wolfman Jack, Michael Faraday, Lord Byron, and the band Metallica appear between the same covers.

# 1

## A World without Science

*"Any sufficiently advanced technology is indistinguishable from magic."*

*—Arthur C. Clarke*

In the early 1800s, the Danish curator and archaeologist Christian Jürgensen Thomsen hit on a novel idea for classifying prehistory artifacts. By dividing them into three categories—Stone Age, Bronze Age, and Iron Age—he was able to make some sense of his museum's collection and shed light on civilizations long vanished. What he had done, of course, was create a technological time line with each of the three classifications defined not only by materials, but also by technical skill sets and accumulated knowledge base. Although modified over the years, Thomsen's three-age system has more or less withstood the test of time.

A hundred years later, F. Scott Fitzgerald would puckishly coin the phrase Jazz Age to define the gaudy up-tempo spree of music, money, bathtub gin, and flappers that defined the decade-long party following World War I. Today we're told we live in the Digital Age, Wireless Age, and Portable Age, though we have yet to come up with a suitable name for the current convergence of all three.

There is a certain appeal in measuring history against the steady advance of science and technology. For one thing, it provides a welcome relief from the tame textbook pageants of politics, personalities, wars, and dates or the sour revisionist history in which flaws overshadow accomplishments currently in favor. And, with few exceptions, science and technology tend to progress in an orderly, logical manner.[1] The time lines are remarkably clear, even in the ancient world.

As far back as 600 BC, Thales of Miletus was already exploring the mysteries of nature. Known as one of the Seven Wise Men of ancient Greece and the father of modern mathematics, Thales left no writing. All that exists of his work are scattered anecdotes from Plato and Aristotle. But even this anecdotal evidence shows the first tenuous, unsteady steps of scientific thought.

In a legend that uncannily parallels the stereotype of the modern absent minded professor, Thales was said to have tumbled into a well (in another version it's a ditch) while contemplating the stars. And during a military campaign against Persia, he supposedly diverted the Halys River (the present-day Kızıl Irmak River in Turkey) by ordering a channel to be dug that allowed a bridge to be built. While in Egypt, Thales was also reputed to have cleverly worked out a way of calculating the height of the pyramids by measuring their shadows on the ground at the time of day a man's shadow is equal to his actual height. A neat trick indeed!

Thales' methodology was simple observation and reason. It was not science in the modern sense of hypothesis and experiment, but rather science based on what was immediately observable. Still, it was

---

1 One significant and troublesome exception is the Baghdad battery. See the appendix for a history of the controversy.

founded on logic and was a significant break from the received wisdom of religious myths that permeated ancient thought. For Thales, nature was neither random nor subject to the whim of the gods. This was a major step forward to be sure, but painfully inadequate when it came to understanding complex natural processes and those things either too large or too small to be seen clearly. Even Thales faced insurmountable problems when it came to studying simple electrical and magnetic phenomena.

For instance, amber, fossilized tree resin, was prized in the ancient world particularly among the Greeks, who called it *ēlektron*—Greek for "gold"—to describe its color, although the term was also used to describe silver. Imported from Burma and traded throughout Europe as early as 1600 BC, it was used in quantity for burial ceremonies by the Greek warrior kings.

Among its more interesting properties, amber could be rubbed to create an electrostatic charge that attracted small pieces of straw, wheat chaff, and thin scraps of copper or iron. Even the most careful observation alone could not reveal the truth that rubbing transferred negatively charged electrons from the cloth or finger to the amber's dry surface, which then attracted the positively charged scraps of wheat. The phenomenon, which today we call "triboelectricity" (from the Greek word *tribō* meaning "to rub") states that two materials exchange electrons when they come in contact with each other. In the process, they form a bond as charges move from one material to the other. When the contact is broken, some atoms keep an extra electron. For instance, when glass is rubbed with wool, the wool acquires electrons and becomes negative, while the glass gives up electrons to become positively charged. By applying a very generous definition—the ability to maintain an electrical charge—amber could also be called the first battery.

Lodestones—naturally occurring magnets—were also problematic for the ancients, including Thales. How do you explain by observation and logic what lies behind a clearly observed, but wholly unlikely, phenomenon: a rock capable of moving metallic objects? According to Aristotle, Thales believed a magnet attracted iron because it had a soul.

The Greeks, and later the Romans, sought to understand the world through observation and the application of logic as opposed to modern and quantifiable methodologies of theory and experiment. For instance, Aristotle, who is credited with the creation of formal logic, held beliefs regarding the formation of metals that were closer to alchemy than science.

Such was the state of science for centuries; when observation and logic failed, myth and magic filled in the gaps. Still, the idea that nature could be known solely through simple observation and the application of logic became central to European scientific thought and persisted as late as the 1600s. As recently as the early 1900s, Aristotle's decidedly vague "fifth element"—*aether*—was still a cause for debate among serious scientists of the day, including Albert Einstein.

The more stubborn myths persisted, echoing through the texts, lending credence to unsubstantiated, often incredible claims. The Roman naturalist Pliny the Elder, the master compiler of nature, included myths and fables alongside his own firsthand observations. In his immense *Historia Naturalis,* the unicorn is given the same credible treatment as the lion. Without a reliable way to verify the stories that came to him, he dutifully recorded folktales and legends that seem outrageous by modern standards. Among them, that ". . . near the River Indus there are two mountains, one of which attracts iron and one repels it. A man with iron nails in his shoes cannot raise his feet from the one or put them down on the other." And why not? Pliny, who was no doubt aware of mysterious magnetic forces, would require only the slightest nudge of imagination to believe those same unseen powers capable of wondrous feats in a faraway land.

To be fair, Pliny was not alone in recording magnetic myths. Farfetched claims regarding magnets were widely circulated. As with Pliny, these often took the form of tales from foreign lands, as if distance suspended natural laws along with the ability to confirm through firsthand observation. In one technological fable that would mutate and endure for centuries, the ancient architect Timochares began to erect a vaulted roof of lodestones in the Temple of Arsinoe at Alexandria so that an iron statue of the queen could be suspended in midair as if by magic. In variations of the same myth, magnets were used to suspend a statue of Mohammed in a mosque while the Venerable Bede, the seventh-century Anglo-Saxon Benedictine monk and author of *The Ecclesiastical History of the English People,* wrote that the horse of Bellerophon—Pegasus, which weighed 5,000 pounds—was levitated by the use of magnets on the island of Rhodes. In China, there were legends of fortresses and tombs made with gates of magnetic stone that acted as a security system by attracting metallic weapons and armor.

STATUE OF PEGASUS
SUSPENDED BY MAGNETS

It is difficult to reconcile the great thinkers—who so keenly and critically mapped the human spirit—giving themselves over to fairy tales. Experimentation in the ancient world, what little of it existed, was the domain of artisans seeking advantage over the competition with closely guarded "trade secrets," early engineers working with well-understood materials and some basic medicine. There were also the alchemists pursuing their futile goals of riches and immortality. In this way, those things of immediate and obvious value or use did progress in the ancient world. It was easy to see the motivation in creating a new soap or beautiful glass beads or the civic benefit of moving relatively large amounts of water through pipes to a thirsty and dirty population.

However, those phenomena that could not be held in the hand or promise immediate benefit remained in the province of philosophy, myth, and religion. What did a magnet or electrostatic charge offer beyond wonder and mystery?

Magnetism took the lead in what little scientific exploration there was of these phenomena in the ancient world. A lodestone could be held in the hand. Its effects were easily seen and even repeated at will, making it a good candidate for study. On the other hand, electricity could only be known as fleeting shocks of electrostatic charges, mysterious, nearly instantaneous, and singularly difficult to study. The torpedo fish, lightning, and electrostatic charges deposited on pieces of amber were all electrical in nature, though the ancients had no way of definitively judging them as the same elemental force. In a world where reality's boundaries were defined by what could be seen, touched, tasted, smelled, and heard, even the most basic understanding of electricity was not only highly problematic, but also ripe territory for myth.

The study of electricity and magnets seemed to creep along for centuries. The Roman poet Lucretius, who sought to elevate reason over superstition, described the power of a magnet in verse in his work *De rerum natura* (*On the Nature of Things*).

St. Augustine in *De civitate dei* (*City of God*) mentions the magnet and its ability to hold a series of iron rings together. "When I first saw it," he wrote,

> I was thunderstruck, for I saw an iron ring attracted and suspended by a stone; and then, as if it had communicated its own property to the iron it attracted, and had made it a substance like itself, this ring was put near another and lifted it up, and as the first ring clung to the magnet, so did the second ring to the first . . . Who would not be amazed at this virtue of the stone, subsisting as it does, not only in itself, but transmitted through so many suspended rings and binding them together by invisible links.

Meanwhile, the myths took root and continued to flourish, expanding as the centuries passed. Tales of lodestones circulated throughout Europe, spread by traders, charlatans, and philosophers. There were lodestones reputed to have the power of discovering thieves and rendering the inhabitants of a house blind. There were lodestones that would absorb iron without adding to their own weight. Lodestones ground up into a powder or held against the flesh with a poultice were touted as cures for colic, insanity, even wounds. It was common wisdom that a lodestone would lose its magnetic power when placed near a diamond or rubbed with garlic and then, miraculously, regain it in full if dipped in the blood of a goat.

Of course, artists could not resist using the mysterious unseen forces—nearly as good as fate, coincidence, or the whim of conflicted gods—to move a story along. Edmund Spenser uses magnetism as a plot device when he describes a magnetic cliff drawing a ship to it in his sixteenth-century epic poem, *The Faerie Queene*:

> *On th'other side an hideous Rocke is pight,*
> *Of mightie Magnes stone, whose craggie clift*
> *Depending from on high, dreadfull to sight,*

*Ouer the waues his rugged armes doth lift,*
*And threatneth downe to throw his ragged rift*
*On who so commeth nigh; yet nigh it drawes*
*All passengers, that none from it can shift:*
*For whiles they fly that Gulfes deuouring iawes,*
*They on this rock are rent, and sunck in helplesse wawes.*

Not much new about the magnet was discovered until around the eleventh century when references to its value in navigation as a primary component in compasses began to appear, at first in Asia and then in Europe. Suddenly, magnets were more than mystifying curiosities; they could perform a practical, even vital, task—guide ships.

Then in the thirteenth century came an unlikely exploration of the magnet. Pierre de Maricourt, called Petrus Peregrinus (or Peter the Pilgrim, a title that indicated he had visited the Holy Land during the Crusades), was in the engineering corps of the French army during the siege of Lucera in southern Italy, where he worked on fortifications and constructed catapults for bombarding the city. A physician with a minimal amount of technical ability, at some point during the siege he hit on the idea of a perpetual motion machine powered by magnets. The machine would, Peregrinus envisioned, turn a small sphere indefinitely using the attractive forces of magnetism.

During the summer of 1269, he put his thoughts into a letter addressed to his close friend Sigerus de Foucaucourt. Rather than simply describe his machine, Peregrinus first set out to describe lodestones in detail, listing attribute after attribute in an orderly fashion. Although his perpetual motion machine was doomed to failure, the first section of the letter is a landmark of inductive reasoning and magnetic science.

"Out of affection for you, I will write in a simple style about things entirely unknown to the ordinary individual," he wrote. "But the things that are hidden from the multitude will become clear to astrologers and students of nature and will constitute their delight as they will also be of great help to those that are old and more learned."

Here was the magnet stripped of speculation, myth, and even poetry. "The disclosing of the hidden properties of this stone is like the art of the sculptor by which he brings figures and seals into existence," he related. "Although I may call the matters about which you inquire evident and of inestimable value, they are considered by common folks to be illusions and mere creations of the imagination."

Soon, hand-transcribed copies of the letter, which became known as *Epistola Petri Peregrini de Maricourt ad Sygerum de Foucaucourt, militem, de magnet* (Letter of Peter Peregrinus of Maricourt, to Sigerus of Foucaucourt, Soldier, concerning the Magnet) began to circulate.

In the pre–printing press age, the vast majority of what little scientific research was undertaken was shared not through scholarly journals or books, but in letters that slowly crisscrossed Europe among a small group of friends and like-minded individuals. In an era when a single book could cost as much as a large tract of land and moveable type was still more than two centuries away, this form of epistolary science was woefully limiting. Nevertheless, Peregrinus's letter was reproduced and referenced in numerous volumes over the years. It crops up in the Franciscan friar Roger Bacon's masterwork overview of science, *Opus majus (Great Work)*, written in secret at the request of Pope Clement IV. Bacon, also known as "Doctor Mirabilis" (Wonderful Teacher), would have recognized Peregrinus's methodology as close to the brand of empiricism he had begun to practice at Oxford and fit into his own category of *scientia experimentalis* (experimental science). Bacon's investigations would eventually cause him to run afoul of the Church. Late in life he found himself convicted of *novitates suspectas* (suspect innovations) and placed under house arrest for more than a decade.

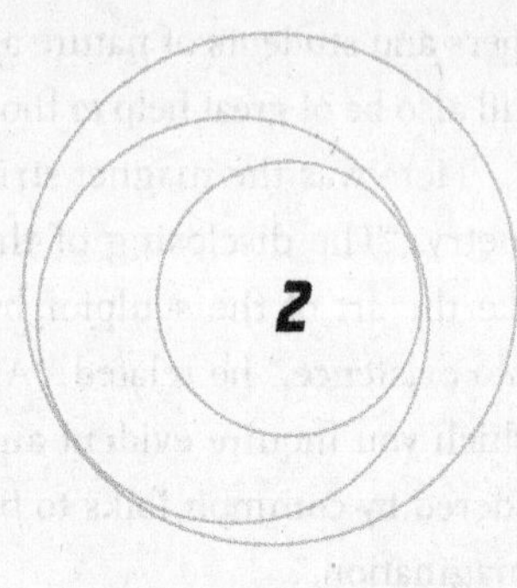

# 2

## *The Death of Superstition*

*"I've found out so much about electricity that I've reached the point where I understand nothing and can explain nothing."*

*—Pieter van Musschenbroek*

In the mid-fifteenth century, the combination of the printing press and faster, safer travel meant that ideas could be shared more readily and among wider audiences. The concept of science stripped of myth began to catch on. However, even as other branches of more practical science, such as medicine, flourished, the study of electricity and magnetism remained nearly stalled. That is, until the Elizabethan physician William Gilbert turned his attention to magnetism. A member of a panel that advised on the queen's health, Gilbert also had a bustling medical practice in London and was a member of the Royal College of Physicians. There could hardly have been a more credible investigator.

Starting his study with amber, as the Greeks had done, he named its attractive power *vis electricia* in Latin—coining a new word—"electricity."

Gilbert's book, popularly called *De Magnete* (The full title is *On the Lodestone and the Magnetic Bodies and on the Great Magnet the Earth*), was published in Latin in 1600 and set the stage for scientific experiment far beyond the study of magnetism and electricity. Widely read and discussed, it presented and proposed nothing less than a new way of doing science.

Others were also experimenting by then and were documenting their results in private letters and pamphlets—though not as exhaustively or meticulously as Gilbert. In 1576, for example, Robert Norman, an instrument maker in Bristol, published a small pamphlet called the *Newe Attractive*. Indeed, Gilbert even duplicates one of Norman's experiments in *De Magnete*. Setting up his methodology at the beginning, Norman wrote, "I meane not to use barely tedious Conjectures or imaginations; but briefly as I may, to passe it over, grounding my Arguments onely uppon experience, reason and demonstration which are the grounds of the Artes . . ." And in Italy, Girolamo Cardano, a doctor, mathematician, and astrologer, published a book called *De Subtilitate Rerum* (*The Subtlety of Things*) that drew the distinction between magnetic and electrical properties.

So then, why do Gilbert and his *De Magnete* get all the praise? First and foremost, his study was the most exhaustive at the time. Not only did he seek out anything he could lay his hands on regarding magnets, but like a good scientist, he duplicated experiments of others to verify the results, tested theories, and added a long list of his own experiments to the mix. Second, unlike the much-neglected Robert Norman, he was in London. Sudden outbreaks of the plague during the warmer months, the emptying of chamber pots carelessly out of windows, and the heads of criminals adorning the Tower Bridge aside, London was a major European metropolis of 75,000 or more and a hub of trade, new ideas, and culture. In Gilbert's London, Shakespeare's talents were in full bloom with plays like *Hamlet* and *Julius Caesar*. And, too, Gilbert was well connected. More than 400 years ago, his book created the right kind of buzz while his position provided ample credibility.

Like Peregrinus before him, Gilbert performed experiment after experiment, carefully detailing the results in writing and only recording what he could verify and repeat. And, almost as important, Gilbert set out to debunk the myths. In Gilbert's view, as with science today, verifiable results trumped myth, no matter how often that myth was repeated. "But lest the story of the loadstone should be *jejune* and too brief, to this one sole property then known were appended certain figments and falsehoods which in the early time no less than nowadays were precocious sciolists and copyists dealt out to mankind to be swallowed," Gilbert wrote.

> . . . The like of this is found in Pliny and in Ptolemy's *Quadriparatitum*; and errors have steadily been spread abroad and accepted—even as evil and noxious plants ever have the most luxuriant growth—down to our day, being propagated in the writings of many authors who, to the end that their volumes might grow to the desired bulk, do write and copy all sorts about ever so many things of which they know naught for certain in light of experience.

Gilbert isn't hesitant to name names, challenging some of the most revered minds in history with the certainty of his experiments.

> Caelius Calcagninius in his *Relations* says that a magnet pickled with salt of the sucking-fish has the power of picking up a piece of gold from the bottom of the deepest well. In such-like follies and fables do philosophers of the vulgar sort take delight; with

> such-like do they cram readers a-hungered for things abstruse, and every ignorant gaper for nonsense.

This was tough stuff for the era—the Elizabethan version of talk radio or departmental brawls in academia.

Remarkably, during the course of his investigation, Gilbert conceived and constructed what is generally believed to be the first electrical device. He called it the *versorium* (turnaround in Latin) and used it for detecting the presence of static electricity. Very simply constructed, the versorium was little more than a metallic needle that pivoted freely on a pedestal. Looking very much like a compass, it could detect the presence of electrical charges from a short distance.

Gilbert also set out a new way to write about science, eliminating all unnecessary prose.

> Nor have we brought into this work any graces of rhetoric, any verbal ornateness, but have aimed simply at treating knotty questions about which little is known in such a style and in such terms as are needed to make what is said clearly intelligible. Therefore we sometimes employ words new and unheard of, not as alchemists are wont to do in order to veil things with a pedantic terminology and to make them dark and obscure, but in order that hidden things which have no name and that have never come into notice, may be plainly and fully published.

Among Gilbert's discoveries was a detailed differentiation between amber—which he called "electrics"—and magnets. "A loadstone lifts great weights; a strong one weighing two ounces lifts half an ounce or one ounce." And on and on goes Gilbert in a decidedly simple style—even translated from the original Latin. By the book's end, he has coined the word "electricity," named the corresponding points of the globe "north pole" and "south pole," differentiated mass from weight, discovered the effect of heat upon a magnetic body, and explained the earth in terms of a celestial magnet. In all, Gilbert's experiments were responsible for more than thirty verifiable new discoveries regarding the magnet and electricity while his methodology set the stage for a new type of scientific investigation.

Perhaps one of Gilbert's most astonishing breakthroughs was his exploration of amber's electrostatic properties. For centuries it had been thought that the secret to amber's attractive powers resided in the way it grew warm when rubbed. Drawing on his experimentation, Gilbert surmised that when amber was rubbed, there was a transfer of "effluvium" to the smooth surface, and that it was this unseen substance that attracted other materials. Of course, he could not have known that the nature of the charge was a transfer of electrons—which would not be identified with certainty until nearly 300 years later, in 1897—but it was an amazingly close conclusion.

To the modern reader, Gilbert's scientific insights are simplistic and even tedious, but to those in his time, they were revelatory. Testing a hypothesis using the most rigorous standards possible was a new concept. Not only did Gilbert present his material clearly and without embellishments, but he set out his methodology free from myths, fables, speculation, and the conveniently distant lands of marvelous tales. For many historians, *De Magnete* marks the end of the Aristotelian reign in science and a dead end for the ancient philosophers known as the Peripatetics, who talked and walked as they applied their logic to life's problems.

Gilbert used logic to be sure, but he applied it systematically to his workshop experiments through inductive reasoning. Building his conclusions through experiment after experiment, he amassed a huge body of data in order to reach those proofs. In the words of the historian of science Park Benjamin, " . . . he, first of all men, systematically replaced the great doctrine of words by the great doctrine of works."

*De Magnete* marked the beginning of modern science, opening the door for Galileo and Isaac Newton. In particular, Newton took to experimentation with sometimes frightening gusto—at one point staring at the sun with one eye to study the afterimages, nearly blinding himself in the process. In another series of experiments, he inserted various instruments around his eye, including a bodkin (an ivory toothpick) as well as his finger to change the shape of his eyeball. "I push a bodkin betwixt my eye and the bone as near to the backside of my eye with the end of it there appear several white dark and coloured circles," he wrote of one experiment.

In his most famous experiment, Galileo, who had called *De Magnete* "great to a degree that it is enviable"—took a somewhat less horrific approach than Newton. Noticing that both large and small hailstones hit the ground simultaneously, he realized there could be only two conclusions: that the larger stones fell more rapidly and had to begin their downward journey from a higher altitude or the counterintuitive explanation that all objects, regardless of weight, fell at the same rate of speed. To determine which of the two options was true, he dropped two objects of different weights off the Leaning Tower of Pisa simultaneously with the result enshrined forever in elementary science textbooks.

Gilbert's book and its conclusions traveled quickly. As early as 1602, he mentions letters from Italy and within a few years a translation was already known in China. Within a few more years it was translated from the original Latin (commonly used for scientific texts) into English.

"To you alone, true philosophers, ingenuous minds, who not only in books but in things themselves look for knowledge, have I dedicated these foundations of magnetic science—a new style of philosophizing," Gilbert wrote in *De Magnete.*

THE BEAUTY OF GILBERT'S EFFORTS—indeed, all scientific work—was that nearly anyone with time and resources could duplicate any of his experiments and achieve pretty much the same results. Unlike the dubious work of alchemists or innovations by tradesmen, which were largely conducted in secret, Gilbert's brand of science was freely shared and open to challenges. A better theory backed up by a credible experiment could displace even the most fundamental of Gilbert's conclusions.

The scientific method would even have a profound effect on alchemy. By the time *De Magnete* was published, the secretive endeavor, which uncomfortably merged the technical and the mystical, had already moved beyond its traditional wasted efforts of transmutation or eternal life toward legitimate medicine. Gilbert and the scientific revolution of the seventeenth century served to push it even further away from magic and mysticism toward experimentation and respectability. In fact, the modern word "chemistry," which came into use in the 1600s, was first used to describe both alchemy and iatrochemistry, though it would take nearly another century to encompass its current scope of study. Notably, the first English use of the word "laboratory" can be traced to 1605.

Science, along with its potential for improving mankind's lot, also slowly began to seep into the general public's consciousness. Francis Bacon penned what may be called the first science fiction book with a work titled *New Atlantis*. Published in 1627, a year after his death, it described a utopian society founded on scientific principles where citizens enjoyed the benefits of such miraculous inventions as the telephone and flying machines. Of course, it was set in a distant land populated by clever people, a plot device that echoes the fantastic fables of magnetic mountains and suspended statuary.

The way the entire universe was perceived was also changing. It was no longer mysterious and unknowable, but, rather, governed by natural laws that acted in a consistent manner. Even more remarkable, those laws could be known and understood by man. It was an exciting prospect, inspiring Alexander Pope to pen the lines

*Nature and Nature's laws*
*lay hid in night;*

*God said, Let Newton be!*
*and all was light.*

The universe, as Newton and Galileo and others came to see it, was very much like a machine that could be puzzled out by experiment and logic. "We are not to imagine or suppose, but to discover what nature does or may be made to do," Francis Bacon wrote, more than just implying that knowledge would bring enormous, unforeseen benefits to those willing to make the effort.

"I do not know what I may appear to the world, but to myself I seem to have been only a boy playing on the sea-shore, and diverting myself in now and then finding a smooth pebble or a prettier shell than ordinary, whilst the great ocean of truth lay all undiscovered before me," Newton famously wrote, imagining an immense territory awaiting exploration, and perhaps even drawing a parallel to a time when European ships had first begun to sail beyond distant horizons in search of new lands.

SMALL GROUPS BEGAN TO FORM to share information and resources, even as attacks on fallacies and myths continued with varying degrees of success. The rise of Puritanism in England placed science under the domination of the Church while quasi-scientific cults began to form, such as the Rosicrucians in Germany, whose members believed, among other things, that magnets could pull diseases from the body.

By 1660, with the death of Oliver Cromwell and the decline of Puritanism, a loosely organized band of twenty-one amateur scientists who had gone by the name "The Invisible College" successfully petitioned King Charles II, and the Royal Society was formed for the purpose of ". . . promoting of experimental philosophy." Overnight, "experimental philosophy"—as science was called at the time—was not only in fashion, but was also a respectable enterprise. Those in the upper tiers of society were soon expected to have at least a rudimentary grasp of the latest scientific theories, discoveries, and experiments in much the same way they were expected to know Plutarch, music, and art.

Landed gentry, noblemen, and successful tradesmen with social ambitions who possessed the time and resources enthusiastically took to science, reading the details of the latest discoveries, supporting research, attending lectures, and even setting up small laboratories of their own to conduct experiments. Not surprisingly, with the popularity of science on the rise among the aristocracy, interest grew among the less fortunate classes. Ben Jonson—a friend of Bacon's—wrote a popular comic play called *The Magnetick Lady or Humours Reconcil'd,* featuring Lady Lodestone as the lead character. Sir Isaac Newton, an intensely private and often paranoid man, paradoxically became something of a pop culture icon of the day with his likeness adorning an endless stream of portraits and commemorative medals.

A kind of cottage industry began to flourish with itinerant men of science offering demonstrations for small towns and villages. Charlatans and natural philosophers of dubious merit soon took to presenting demonstrations in parlors and lecture halls, many of them making up theories as they went along.

To feed the curiosity of a public eager for the latest news of scientific discovery, publications also multiplied, ranging from the serious science of the Royal Society's *Philosophical Transactions* to popular titles such as *Sir Isaac Newton's Philosophy Explained for the Ladies.* In the mid-1600s, a London physician named Thomas Brown published *Pseudodoxia Epidemica* or *Enquiries into Vulgar and Common Errors.* Like Gilbert, he intended to discredit unproven and untested beliefs. Although an instant bestseller, quickly selling out multiple printings, the book is generally seen as a mishmash of solid experimentation and confusing explanations, though he did manage to coin several new words. And Robert Boyle, the father of modern chemistry, published *Experiments and Notes about the Mechanical Origine or Production of Electricity* in 1675, generally seen as the first book dedicated entirely to electricity.

. . .

THE MACROSCOPIC SCIENCES—BASIC PHYSICS, botany, astronomy, and even anatomy—advanced far ahead of the studies of chemistry and electricity. For scientists laboring in the seventeenth and into the eighteenth century, the world of molecules and electrons was so closed off they might just as well have existed on a different planet—or beyond the horizon of that vast ocean of knowledge imagined by Newton. Electricity, in particular, was extremely difficult to study. Although definitively separated from the effects of magnets by Gilbert, the sudden burst of an electrostatic spark was a poor substitute for an extended flow of current.

Not surprisingly, the first machine built to produce a steady electrical charge bore a striking resemblance—at least in principle—to the ancient technique of rubbing amber, though its inventor arrived at it in a somewhat roundabout manner. Otto von Guericke from Magdeburg, Germany, was born into a prominent family and duly educated in the style provided by prominent families of the day. Beginning his studies at the University of Leipzig, he moved on to Jena for law and finally engineering in Holland, at Leyden. Back in Magdeburg, he saw the tiny community through the devastating armed conflicts of the Thirty Years' War. Somewhere along the way, he found time to dabble in science.

By 1654, Guericke invented an air pump that gained him some notoriety, and then, following in Gilbert's steps a few years later, he built what he believed was a scale model of the earth. Coating the inside of a glass sphere—about the size of an infant's head, according to his description—with sulfur and minerals, he heated it and then broke the glass, leaving a perfectly round sulfur sphere. Taking the sphere, he placed it in a machine that turned it against a pair of leather pads in an effort to simulate planetary powers. Apparently he had accepted Gilbert's theory that electricity and gravity were linked, but he unwisely rejected the idea of the earth as a giant magnet. Despite his poor assumptions, what he had done was give the

VON GUERICKE GENERATOR

hollow sulfur ball an electrostatic charge. Once charged, the sphere attracted light objects—just as the amber had in ancient times—such as feathers and bits of cloth while rejecting other substances.

In just a few years, the Royal Society was conducting its own experiments using a similar machine. Francis Hauksbee, Newton's longtime assistant turned London instrument maker and the Royal Society's chief "experimentalist," along with Christian August Hausen, a professor of mathematics in Leipzig, soon improved on Guericke's machine, replacing the sulfur sphere with one of glass.

Finally, with a reliable way of generating and briefly holding an electrical charge, experimenters began probing the nature of electricity. Lectures and public experiments in the major cities of Europe cropped up along with a demand for similar devices among the well-to-do. Purchasing the electrostatic generators out of curiosity or for entertainment, amateur scientists duplicated the experiments they had seen in lecture halls or read about in the scientific journals of the day. George Matthias Bose, a German professor, began experimenting with electrical devices, attempting to increase the discharge by adding a series of rotating glass globes, using three globes varying in size from ten to eighteen inches. So powerful was the combined charge, he claimed, that blood flowing from the vein of an electrified person seemed to glow.

And in Erfurt, Germany, Andrew Gordon, a Scottish Benedictine monk, was also busily experimenting. In one landmark experiment, he arranged two metal gongs with a metal ball suspended between them on a silk thread. When one gong was electrified via an electrostatic machine and the other grounded, the ball swung toward the electrified gong, striking it, then swung to the other, which was grounded, and back again. Although largely forgotten to history, Gordon's simple device was the first known instance of anyone converting electrical energy to mechanical energy. Eventually his invention became known as the "German chimes." However, a few years later when Benjamin Franklin unwisely used a lightning rod to pull an electrostatic charge down into his parlor from approaching storm clouds to ring bells, they quickly became known as "Franklin chimes."

Gordon is also credited with creating the first electric motor. An ingenious device, it was based on the same principle as the ancient steam engine known as the aelopile of Hero, invented by the Alexandrian mathematician around 200 BC, which released steam through two openings on opposite sides that sent the sphere spinning on a spitlike device. Called the "electric whirl," Gordon's motor was a metallic star that pivoted at its center. When subjected to an electrical charge at the points, it spun.

Electricity was also big news in the eighteenth century, particularly in England where the *British Magazine*, the *Universal Magazine*, the *London Magazine*, and other popular publications regularly reported news of the latest electrical experiments. However, even the serious experimenters could not resist a bit of joking. According to reliable accounts, Bose would electrify metal dinner plates and force an electrical charge through a coin held between a volunteer's teeth or through a group of people holding hands.

Then in 1706, Francis Hauksbee the Elder offered up another tool for electrical experimentation to the Royal Society—a simple glass tube some thirty inches long. When rubbed with a piece of cloth, it also held an electrical charge. Public demonstrations of the tube caused a sensation. Using thin slivers of brass, threads, and even his own hair, rather than wheat chaff, Hauksbee managed an entire series of crowd-pleasing public demonstrations.

Unlike Guericke's electrostatic machine, the glass tube was not only a reliable way of storing a charge, albeit very briefly, but was also simple to charge and inexpensive to manufacture. Virtually anyone with an interest could acquire one of the tubes, opening the doors of experimentation to an even greater number of amateur scientists. One of the more unlikely experimentalists who adopted the glass tube was Stephen Gray from Canterbury. A dyer by profession, he was also an avid amateur scientist. Over the years his enthusiasm for science led him from astronomy and optics to the supernatural and finally electricity. Eventually pensioned off, he was admitted to an institution in London known as the Charterhouse, a monastery closed during the Reformation and converted into a home for retirees and a day school for poor boys.

Gratefully freed from a business for which he seemed to have little enthusiasm and safely ensconced in the institution, he indulged his passion for science and experiments. Unable to afford the mechanical electrostatic generator, Gray and two colleagues began with a series of experiments using a glass tube, demonstrating that hair, feathers, silk, and organic material, such as ox guts and even a live chicken could act as conductors to transmit an electrical charge. At one point they suspended a "charity boy," no doubt enlisted from the Charterhouse day school, by silk threads. Then, by touching the child's feet with a charged glass tube, Gray and his friends found brass slivers were attracted to the youth's face. Electricity, it seemed, was able to travel through a living human being as easily as through a chicken or ox guts. In another experiment Gray managed to send an electrical charge from his glass tube nearly 900 feet through a length of line.

Perhaps Gray sought to do for electricity what Gilbert had done for the magnet. Conducting experiment after experiment, he compiled lists of which materials would and would not carry electrical current. Although valuable at the time, his research was not as scrupulous as Gilbert's. Possibly influenced by his former profession as a dyer, he concluded that color played a role in conductivity. According to Gray, red, orange, and yellow were better conductors than blue, green, or purple.

However, in what would evolve into scientific tradition, Gray's experiments were soon repeated and built upon. Key to this effort was Charles-François de Cisternay Du Fay in France. Independently wealthy, brilliant, and a member of the Paris Academy of Science, he was everything a natural philosopher should be in his day, most of all meticulous. A former army officer who had lost a leg in combat, he applied himself to science with a tireless and rigorous discipline. Du Fay had a nimble mind, mastering chemistry, anatomy, botany, geometry, astronomy, mechanical engineering, and antiquities during his short life. Finally, turning his attention to electricity, he repeated Gray's experiments with conductors where he found and corrected the flaws, before beginning his own, more complex experiments. Du Fay also discovered that electricity was released in two forms—which he called "resinous" (for negative) and "virtuous" (for positive).

Despite his exhaustive experiments, Du Fay felt that he had not puzzled out electricity's place in nature. Still, the experiments had yielded enough, so that he no doubt glimpsed a hint of its elusive makeup. In his last memoir, dated 1737, he wrote, "Electricity is a quality universally expanded in all the matter we know, and which influences the mechanism of the universe far more than we think."

THEN CAME THE BREAKTHROUGH—SORT of. In 1745 Ewald Jürgen von Kleist, the dean of the cathedral chapter at Kammin in Pomerania, had the idea that storing electricity might be a good idea to further his experiments. Beginning with Du Fay's finding that water has a natural affinity for electricity, he set to work. Barely a month after beginning his research efforts to store an electrical charge, he related his news of just such a device to the physicist and doctor J. N. Lieberkühn and a few others by letter. "If a nail, strong wire, etc., is introduced into a narrow-necked little medicine bottle and electrified, especially powerful effects follow. The glass must be very dry and warm. Everything works better if a little mercury or alcohol is placed inside," Kleist wrote. "The flare appears on the little bottle as soon as it is removed from the machine, and I have been able to take over sixty paces around the room by the light of this little burning instrument."

The instrument Kleist stumbled upon was a condenser—a kind of battery—that he charged through the nail point with an electrostatic generator similar to Guericke's machine. The free electricity inside created a corona discharge emitting a dim light. Unfortunately, none of those who received his letters was able to repeat the experiment successfully. Kleist either neglected to mention or failed to notice that the bottle's exterior must be grounded—by holding it—while charging.[2]

Next came Pieter van Musschenbroek, a professor at the University of Leyden (today Leiden). The son of a prominent instrument maker who turned out telescopes, microscopes, and other devices, and the protégé of Willem Jakob 's Gravesande, a mathematician, physicist, and disciple of Newton, Musschenbroek was a talented and meticulous experimentalist.

However, as with so much of the science documented by letter, the particulars surrounding the invention of the first true electrical

2 Kleist was not alone in his efforts. Bose, who had also read Du Fay's accounts of water drawing electricity from electrified glass, had the idea of reversing the experiment and electrifying water, but little came of it. And Gordon, the Scottish monk, also experimented with both electrostatic machines as well as attempting, with little success, to store electricity in a container filled with water.

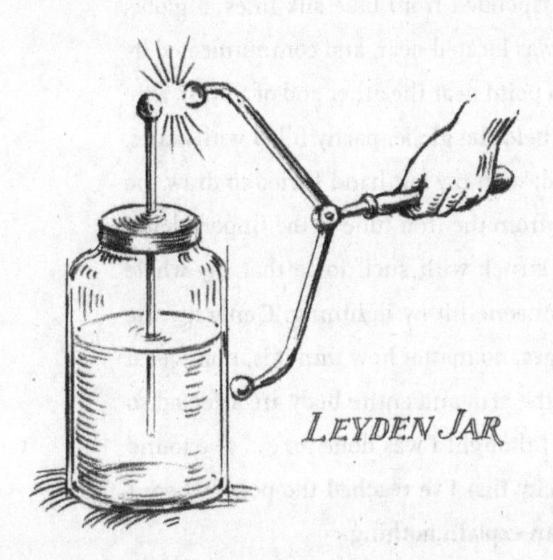


storage device is in question. What is known is that Musschenbroek was trying to duplicate some of the British experiments that produced sparks. In one version of the story, Musschenbroek's friend, a local lawyer by the name of Andreas Cunaeus, was visiting him in the lab. An amateur scientist himself, Cunaeus tried to reproduce Bose's experiment at home. Lacking an assistant, he held the jar while charging it and received a powerful shock. ". . . it knocked the wind out of me for several minutes," he is reputed to have reported. Two days later, Musschenbroek duplicated the lawyer's experiment, substituting a large globe for the small jar, and received an even stronger shock. In another, far more likely version, Musschenbroek created the device himself and may have used Cunaeus as an assistant in the lab.

News of the discovery spread quickly throughout Europe. After reporting some meteorological observations to a colleague at the Paris Academy in a letter dated January 1746, Musschenbroek found himself with a partially filled sheet of paper. Rather than simply end the letter and "waste" the unused space, he wrote:

> As I see this sheet is not completely filled, I would like to tell you about a new but terrible experiment, which I advise you never to try yourself, nor would I, who have experienced it and survived by the grace of God, do it again for all the kingdom of France. I was engaged in displaying the powers of electricity. An iron tube [said

> to be a gun barrel] was suspended from blue silk lines: a globe, rapidly spun and rubbed, was located near, and communicated its electrical power . . . From a point near the other end of a brass wire hung; in my right hand I held the globe, partly filled with water, into which the wire dipped; with my left hand I tried to draw the snapped sparks that jump from the iron tube to the finger; thereupon my right hand was struck with such force that my whole body quivered just like someone hit by lightning. Generally the blow does not break the glass, no matter how thin it is, nor does it knock the hand away; but the arm and entire body are affected so terribly I can't describe it. I thought I was done for . . . I've found out so much about electricity that I've reached the point where I understand nothing and can explain nothing.

What he had created was a condenser or capacitor that stored electrical energy. The way the device worked was simple: the glass jar itself is a nonconductive material, called a dielectric in technical parlance. A layer of metal (conductor) was wrapped around the inside. The person holding it on the outside (later replaced by a second piece of metal) acted as a ground. To charge the jar, Musschenbroek pulled current from an electrostatic machine to a wire protruding from the top, which charged the inner metallic surface. When the outside metallic surface or hand holding the jar was connected via a conductor, an electrical charge was produced as electrons rushed from the inside to the outside.

While essentially a duplicate of Kleist's work, Musschenbroek gets the credit because his more detailed letter allowed others to easily replicate the "terrible experiment." Needless to say, scientists across Europe were soon following suit, replicating the awful and dreadful effects with their own electrically charged jars. In his original letter, the detail-oriented Musschenbroek specified that only German glass be used, but others quickly discovered that almost any type of glass worked just as well, regardless of nationality.

Although neither Musschenbroek, nor anyone else at the time, understood the physics involved, they immediately grasped the value of the device as a new way to study electricity. While the jar sent off only a single jolt of current, it was more powerful and easier to study than the charge produced by any of the other available machinery or electrostatically charged glass rods or globes in use.

Within a very short time Leyden jars (as they soon became known) were created in both France and England, and in due course the effects of electric shocks were reported to include nosebleeds, paralysis, convulsions, and other extreme results. The French physicist and royal electrician Jean-Antoine Nollet (an abbé of a minor order whose ecclesiastical robes made his scientific studies acceptable to the royal court) is credited with introducing the Leyden jar to France. His later experiments included shocking a sparrow, which he turned over for analysis by a surgeon who noticed its insides resembled those of a man struck by lightning.

Nollet also entertained the French king by transmitting a charge through 180 guards ". . . who were all so sensible of it at the same instant that the surprise caused them all to spring at once." In Paris, another experimentalist, Louis-Guillaume Le Monnier, managed to send a shock through a mile-long line of Carthusian monks, each holding on to an iron wire.

Experimenters continued to probe the mysteries of electricity with the jar. During one experiment, Le Monnier tried to determine the speed at which electricity traveled by creating a mile-long circuit. While too rapid to measure, he estimated the speed was at least thirty times the speed of sound. He also experimented with sending current through water. At the Royal Society, experimenters brought a jar to the river and tried to measure its velocity through two miles of water. And during one rather odd piece of research, a French experimenter tested the theory that it was impossible to electrify a eunuch. A castrato was eventually recruited and put in line with two other men. The castrato jumped right along with his two companions.

The jar itself became a subject of experimentation and improvement. The German physicist Daniel Gralath found that connecting several jars in parallel in what he first called an "electrical battery"—appropriating military terminology—increased the power. Nollet discovered that while both the inside and the outside of the jar needed to be dry and clean, any nonoily liquid could replace water and the shape of the vessel didn't matter. Experimenters tested the effects of electricity on vegetables as well as animals. Nollet's experiments led him to conclude that plants subjected to an electrical charge grew faster and that electrified cats lost weight.

At about the same time, serious research was taking place in England at the Royal Society. William Watson, who had begun his career as an apothecary's apprentice with a natural inclination toward botany and who eventually set up his own business before entering the field of science, discovered the electrical circuit. And the aristocratic Henry Cavendish, at best a difficult personality, used the Leyden jar to conduct an unprecedented series of experiments on the conductivity of different metals and the very nature of electricity—the majority of which he didn't publish. In Germany, Johann Heinrich Winkler, a professor of Greek and Latin at the University of Leipzig as well as amateur scientist, made several important discoveries, including that when electricity is given multiple paths to choose from, it invariably chooses the best conductor.

With the ability to store an electrical charge, for even a brief time, it was only a matter of time before someone figured out a way to put the stored energy to use. Ebenezer Kinnersley, a Baptist minister, and Benjamin Franklin's colleague in experimental philosophy, invented a wheel that turned when thimbles along its edge touched wires leading back to a Leyden jar; another version, made of glass, rang chimes.

Hobbyists and serious natural philosophers also tried to measure electrical output. In France, Nollet, still the leading experimenter of his day, created what could be called an "electroscope" that measured the movement of two dangling electrified strings from a Leyden jar projected on a screen. Somewhat similar devices came out of England and featured pith balls that operated on the same principle as Gordon and Franklin's chimes, while other instruments, such as the electrometer, used thin strips of metal foil to measure the strength of a charge.

These ingenious pieces of equipment were designed to measure electric current without a full understanding of what electric current was exactly. Still, as crude and inaccurate as they were, they offered experimenters some way to judge the output of stored energy in a Leyden jar.

AS SCIENTIFIC STUDY BLOOMED IN Europe and news of the Leyden jar spread via letters and scientific journals, America remained a scientific backwater. Despite some growing interest among the intelligentsia, American science had certainly not reached the levels of interest seen in Europe. Experimental philosophy was not in vogue in America as it was in England and France. There were no learned societies, and men of leisure found other diversions. What little innovation took place, even in the large cities of New York, Philadelphia, and Boston, was dominated by tradesmen with little time or inclination for the less than pragmatic study of electricity.

In the summer of 1743, Benjamin Franklin happened to attend Dr. Archibald Spencer's demonstration in Boston, which featured one of Hauksbee's glass rods. A popular traveling lecturer from Edinburgh, Spencer apparently repeated some of Gray's experiments, including dangling a young boy from strings. The allure of science must have been irresistible. As with nearly everything he did, Franklin threw himself into electrical experimentation with gusto. From 1746 until 1752 these efforts consumed him.

A glass tube along with a sampling of the latest scientific literature was soon sent to him by his friend Peter Collinson. A London cloth merchant, agent for Franklin's Library Company of Philadelphia, and a fellow of the Royal Society, Collinson was an ideal conduit into European scientific circles. Although his own field of interest was botany rather than electricity, Collinson was nevertheless an enthusiastic and often effective supporter of Franklin's work.

Within a year, Franklin was writing letters back to Collinson detailing his discoveries. The early letters, although read before the Royal Society, remained unpublished in *Philosophical Transactions*. Soon after receiving his initial glass rods from England, Franklin commissioned additional tubes made locally from plain green glass. It was rumored that he rubbed them with buckskin to create a charge, but he soon switched to an electrostatic machine very similar to Guericke's that was fashioned by a local silversmith.

"I never before was engaged in any study that so totally engrossed my attentions and my time as this has lately done," Franklin wrote to Collinson in 1747. "What with making experiments when I can be alone, and repeating them to my friends and acquaintances who from the novelty of the thing come continually in crowds to see them I have during some months past had little leisure for anything else."

Franklin's fascination also extended to Leyden jars. Working with Kinnersley as his assistant, he set about improving the jar's design. Either he or Kinnersley coated the outside with metallic foil, essentially replacing the user's hand as a cathode that accepted the electrons. Then, in one of his groundbreaking experiments, Franklin set about to learn just how the mysterious force and the jar worked. Where, exactly, did the powerful electric charge—the "subtle fluid"—reside in the jar? After charging a Leyden jar, he began to dissect it—testing each component for an electrical charge, even switching out the water of a charged jar.

What he discovered, and related to Collinson in a letter, was that the charge was created by the sum of the jar's components—the two metallic surfaces divided by the nonconductive glass. The power of the electrical charge was in its movement from one metallic surface to another. What Franklin had done in modern parlance was to reverse engineer the Leyden jar. Once he understood how the components worked, if not the science behind them, it was a simple matter to improve on them.

> Upon this we made what we called an electrical battery consisting of eleven panes of large sash glass armed with thin leaden plates, pasted on each side, placed vertically, and supported at two inches distance on silk cords, with thick hooks of leaden wire, one from each side, standing upright distant from each other, and convenient communications of wire chain, from the giving side of one pane to the receiving side of the other so that the whole might be charged together, and with the same labor as one single pane . . .

However, it was one of Franklin's first discoveries—how pointed bodies are more adept than flat surfaces at attracting and emitting an electrical charge—that would establish his reputation in a way he could not have anticipated. Franklin's initial notion was that a grounded metal rod with a pointed tip might actually dissipate a thunderstorm by drawing the current from the clouds and rendering it harmless. Instead, he found that the rod, while protecting buildings, also seemed to actually attract lightning strikes. Still, as he soon realized, it was one thing to puzzle out natural phenomena in a lab with Leyden jars and quite another to leave the lab and bring science to the world at large.

At the time, lightning was still seen very much as a religious matter, sent down from the heavens by demons or a peevish god exercising divine wrath. For the pious of Franklin's day, there was little doubt that lightning rods were evil in that they suborned God's will. If God wanted a structure struck by lightning, then what right did Franklin and his gadget have to subvert those wishes? That's not to say there wasn't a virtuous way in which to safely avoid strikes. Even in Franklin's time, the accepted way to keep churches—often the highest point in a town or city—safe from lightning strikes was to ring the church bells in the belief that the sound somehow broke up cloud formations, rendering them harmless. Unlike Franklin's ungodly science, this was a "pious remedy" and not likely to offend the Almighty, though it did cause several bell ringers to lose their lives from lightning strikes. Some early consecrations of church bells even included prayers that their sound would " . . . temper the destruction of hail and cyclones and the force of tempests; check hostile thunders and great winds and cast down the spirits of storms and the powers of the air."

Franklin was condemned from pulpits as well as in pamphlets, and when the Boston earthquake (also known as the Cape Ann earthquake) of November 18, 1755, struck, Franklin was blamed by at least one minister for sending lightning into the ground to cause it. Rev. Thomas Prince of Boston's historic Old South Church, where a few years later Samuel Adams would signal the start of the Boston Tea Party, railed against Franklin in a sermon titled "Earthquakes the Works of God and Tokens of His Just Displeasure" that offered up an odd mixture of scientific speculation and religious fervor. "The more Points of Iron are erected round the Earth, to draw the Electrical Substance out of the Air; the more the Earth must need be charged with it," he wrote.

> And therefore it seems worthy of Consideration, Whether any Part of the Earth being fuller of this terrible Substance, may not be more exposed to more Shocking Earthquakes. In Boston are more erected than anywhere else in New England; and Boston seems to be more dreadfully shaken. O! there is no getting out of the mighty Hand of God! If we think to avoid it in the Air, we cannot in the Earth: Yea it may grow more fatal.

Franklin even crossed England's King George III, who believed that the points were unnecessary and perhaps dangerous in that they attracted lightning. He preferred blunt rods and so, blunt rods were what England got. Today, more than 250 years later and backed up by the latest science, it turns out that blunt rods are actually more effective.

Franklin was not the only one to hit on the idea of the lightning rod. A Czech priest by the name of Prokop Divi seems to have come up with the idea independently. However, it was Franklin who garnered the credit along with the criticism.

And there was plenty of criticism. Even as Franklin was condemned from the pulpit at home and the throne in England, his letters and reports sent to the Royal Society received mixed reviews. His paper on the "sameness of lightning with electricity" drew laughs when read at the Royal Society, but Collinson remained a staunch supporter. Unable to publish the majority of Franklin's letters in *Transactions*, he approached Edward Cave, the publisher of *Gentleman's Magazine*, the first British general interest magazine.

The son of a cobbler, Cave had bounced from one unproductive career to the next before hitting on the highly successful idea of *Gentleman's Magazine*. He both lived and worked in the publication's offices at London's St. John's Gate, and it was said that he rarely ventured out. A businessman first and foremost, Cave knew a good thing when he saw it, and one of the good things he had previously seen was Samuel Johnson. The general public was fascinated by news of electrical experiments and the fact that these experiments were conducted in the Colonies only added novelty.

Cave agreed to publish Franklin's letters in a modest book called: *Experiments and observations on electricity, made at Philadelphia in America, by Benjamin Franklin, L.L.D. and F.R.S. To which are added, letters and papers on philosophical subjects. The whole corrected, methodized, improved, and now first collected into one volume . . .*" The price: two shillings and sixpence.

The small book, hardly more than a pamphlet, was soon translated into German, Italian, and French. Some of the experiments and conclusions so offended Nollet at the royal court in France that he at first thought the science conducted by an American colonialist and tradesman was a hoax and took to writing pamphlets countering the American's findings without duplicating the experiments.

Franklin appeared unaffected by Nollet's objections and wrote about it quite sensibly.

> He [Nollet] could not at first believe that such a work came from America, and said it must have been fabricated by his enemies at Paris, to decry his system. Afterwards, having been assur'd that there really existed such a person as Franklin at Philadelphia, which he had doubted, he wrote and published a volume of Letters, chiefly address'd to me, defending his theory, and denying the verity of my experiments, and of the positions deduc'd from them . . . I concluded to let my papers shift for themselves, believing it was better to spend what time I could spare from public business in making new experiments, than in disputing about those already made.

Included in the small book was one of the most famous experiments of all time, intended to prove that lightning was, in fact, electrical in nature.

> Make a small Cross of two light Strips of Cedar, the Arms so long as to reach to the four Corners of a large thin Silk Handkerchief when extended; tie the Corners of the Handkerchief to the Extremities of the Cross, so you have the Body of a Kite; which being properly accommodated with a Tail, Loop and String, will rise in the Air, like those made of Paper; but this being of Silk is fitter to bear the Wet and Wind of a Thunder Gust without tearing. To the Top of the upright Stick of the Cross is to be fixed a very sharp pointed Wire, rising a Foot or more above the Wood. To the End of the Twine, next the Hand, is to be tied a silk Ribbon, and where the Twine and the silk join, a Key may be fastened. This Kite is to be

> raised when a Thunder Gust appears to be coming on, and the Person who holds the String must stand within a Door, or Window, or under some Cover, so that the Silk Ribbon may not be wet; and Care must be taken that the Twine does not touch the Frame of the Door or Window. As soon as any of the Thunder Clouds come over the Kite, the pointed Wire will draw the Electric Fire from them, and the Kite, with all the Twine, will be electrified, and the loose Filaments of the Twine will stand out every Way, and be attracted by an approaching Finger. And when the Rain has wet the Kite and Twine, so that it can conduct the Electric Fire freely, you will find it stream out plentifully from the Key on the Approach of your Knuckle. At this Key the Phial may be charg'd; and from Electric Fire thus obtain'd, Spirits may be kindled, and all the other Electric Experiments be perform'd, which are usually done by the Help of a rubbed Glass Globe or Tube; and thereby the *Sameness* of the Electric Matter with that of Lightning compleatly demonstrated.

Contrary to popular legend, it wasn't the forty-six-year-old Franklin who first conducted the famous "kite experiment," but rather, a Frenchman from Bordeaux, Thomas-François d'Alibard, who had read a somewhat poor translation of the American's proposed experiment and decided to try it himself. In a field just outside of Paris, he constructed a sentry box with a forty-foot iron rod attached to a Leyden jar.

Then, on May 10, 1752, as a thunderstorm approached, he left the sentry box in the care of a former dragoon by the name of Corffier. When lightning struck the rod, the old man panicked and called for help, bringing the local priest and a small group running through the storm to his aid. The cool-headed cleric followed the experiment, taking the older man's place. "I repeated the experiment at least six times in about four minutes in the presence of many persons," the priest wrote. "And every time the experiment lasted the space of a *pater* and an *ave.*"

Proof positive came when the French priest used a conductor to draw off electricity from the jar. D'Alibard could not have asked for a more credible eyewitness than a priest. News of the experiment's success spread quickly and within weeks Franklin was the toast of France, though he had yet to realize it. "Franklin's idea ceases to be a conjecture," says d'Alibard in concluding his report to the French Academy.

A month later, Franklin, who still did not know of the Frenchman's success, tried the experiment himself. Here again, popular mythology conveniently enters the scene. In most pictures, his son, William—who acted as assistant—is portrayed as an eager child. In fact, William, born out of wedlock to a mother who remains a mystery to this day, was an adult at the time of the kite experiment.

As with so much in early science, Franklin's experiment remains clouded with some doubt and, according to some, controversy. A complete account of the experiment was not recorded until a full fifteen years later in a massive two-volume tome titled *The History and Present State of Electricity* authored by the English scientist Joseph Priestley, but edited (and some say largely written) by Franklin. The question remains: why did he wait so long before publishing his results?

Clearly, Franklin had no idea just how dangerous his experiment was or he might have offered something akin to a disclaimer. In Russia, Georg Wilhelm Richman, a German scientist employed by the Tsar, turned himself into toast while attempting to duplicate the kite experiment and entered history books as the first electrical fatality during an experiment. The Tsar immediately banned all electrical experiments.

In France, Franklin was hailed as a scientific hero. The upstart American colonist was not the first to speculate that lightning was electrical. Isaac Newton, among others, held that view, but it was Franklin who proved it. Acclaim, in both the colonies and Europe followed. Harvard presented Franklin with an honorary degree, and

Yale, along with the College of William and Mary, soon followed. The Royal Society, which had once laughed at his theories, presented him with the Copley Medal and membership.

"*The Tatler* [an early colonial magazine intended to ". . . pull off the disguises of cunning, vanity, and affectation" to lead its readers toward better Christian lives] tells us of a girl who was observed to grow suddenly proud, and none could guess the reason till it came to be known that she had got on a pair of new silk garters," Franklin wrote. ". . . I fear I have not so much reason to be proud as the girl had; for a feather in the cap is not so useful a thing, or so serviceable to the wearer, as a pair of good silk garters."

By the time he completed his study of electricity, Franklin had coined the words "charged," "charge," "condense," "discharge," "electrical fire," "electrical shock," and "electrician." He was also the first to use, at least in English, the words "battery," "conductor," "electrify," and he replaced what Du Fay had called "resinous" and "virtuous" with the terms "negative" and "positive."

Franklin, who Immanuel Kant called "the new Prometheus," had no real immediate scientific successor. On the edge of wilderness and far from the Royal Society or the Academy in Paris, America was still a country of pragmatic tradesmen and artisans. Electricity offered no immediate benefit.

In Europe, the situation was much different. The eighteenth century came to a close with natural philosophy firmly established as a respectable interest, at least among those who could afford to indulge their curiosity. No less a figure than Erasmus Darwin, a polymath sage and the grandfather of Charles Darwin, suggested that young ladies attend lectures by natural philosophers in order to improve themselves, presumably in the same way they might learn a musical instrument or the basics of sketching. Electrical demonstrations continued to attract crowds of the scientifically minded and the curious, and the simple design of the electrostatic generators and Leyden jars made them easy to construct for the well-heeled who wished to conduct their own experiments. Mathematics and the basic principles of engineering, which had gotten their start in the lab, now expanded out into the marketplace, particularly in England. Within just a few generations—from the mid-1700s to about 1800—England shifted from a largely agricultural economy into a growing industrial force as the Industrial Revolution began to take hold.

Still, even as knowledge of chemistry, math, and hydraulics blossomed and was duly applied to pragmatic pursuits in commerce, electricity remained a mysterious force without much, if any, utility.

3

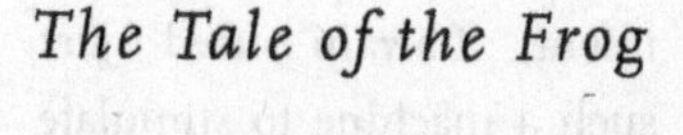

# The Tale of the Frog

*"The language of experiment is more authoritative than any reasoning: facts can destroy our ratiocination—not vice versa."*

*—Alessandro Volta*

When it comes to the history of the battery, we find not one, but three, different versions of Luigi Galvani's 1786 discovery. Even science historians love a good story. Can anyone—except a curmudgeon—deny Newton his falling apple or Archimedes his *Eureka!* moment in the bathtub? As long as the science is solid, then let the stories be told.

In one version of the tale, Galvani was in his lab preparing to sit down for a lunch of tasty frog when he saw his entrée's leg twitch at the touch of a knife. In another and somewhat poignant account, he was dutifully preparing a lunch for his invalid wife when he spotted the same phenomenon. However, in the third and far more probable version—for those who insist on accuracy—the good professor of anatomy and obstetrics in Bologna, Italy, was in his lab preparing a frog for dissection on a metal plate. When he touched the blade of his scalpel to the deceased amphibian, its leg noticeably and quite unexpectedly twitched.

Galvani's first thought was that electrical fluid had somehow jumped several feet from a nearby electrostatic machine to act upon the frog's nerves. Given what was known at the time, this was not an altogether unreasonable theory. After all, just a few years before, he had demonstrated the electrical basis of nerve impulses. In that experiment, Galvani used just such a machine to stimulate movement in a partially dissected frog through the crural nerves that run downward from the lower back to the leg.

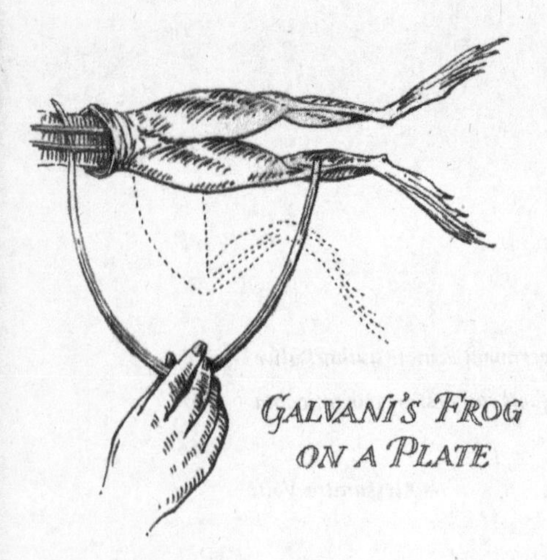
GALVANI'S FROG ON A PLATE

Galvani was also not the first to make such an observation. In 1752, the Dutch biologist and entomologist Jan Swammerdam had witnessed the same phenomenon but gave the effect only a brief mention in his book *Biblia Naturae*. And too, the effects of an electrical jolt on the muscles of animals had been known since earlier experiments with Leyden jars. However, as Galvani had noted in these earlier experiments, it was the nerves and not the muscles that caused contractions when given a jolt of electricity.

What Galvani was witnessing at the touch of the metal blade was an entirely different and more complex phenomenon. Through a series of experiments, he quickly eliminated the electrostatic machine from the equation and soon published a paper that explained the jerk of the frog's leg by a completed circuit through the crural nerve and the leg muscle. The twitching, he reasoned, was the result of electricity accumulated in the muscle traveling through the circuit. According to his 1791 paper, called *De Viribus Electricitatis in Motu Musculari Commentarius* (Commentary on the Effect of Electricity on Muscular Motion), the frog's leg muscle acted like a charged Leyden jar to activate the nerve.

"Perhaps the hypothesis is not absurd and wholly speculative which compares a muscle fibre to something like a small Leyden Jar or to some similar electrical body charged with a two-fold and opposite electricity," Galvani wrote. "And by comparing a nerve in some measure to the conductor of the Jar, in this way one likens the whole muscle, as it were, to a large group of Leyden Jars."

He was, of course, wrong. But where else could the charge have originated? Electricity, already proven to stimulate nerves, must have come from somewhere. And the muscle was as good a candidate as anything else after the only likely external source of a charge—the electrostatic machine—was ruled out. Meanwhile, on the other side of Italy, Professor Alessandro Volta, a creator of scientific instruments, took exception to Galvani's findings, summarily dismissing them as "unbelievable."

Born into an old Lombard family, Volta was first tempted by the Jesuits and then steered by an uncle into law before turning to science. His decision, penned in something of a formal announcement while still a student, was in the form of a poem praising the ideals behind natural philosophy, specifically chemistry and electricity. Notebooks in which the young Volta ruminated on the soul of animals—theorizing that animals also had spiritual powers akin to those of people—could not have bolstered his standing among the Jesuits.

Precociously ambitious, while still a teenager he began an active, though somewhat one-sided, correspondence with the leading natural philosophers in Europe. Announcing his sweeping scientific theories, he sent off letters to Jean-Antoine Nollet, among others, but received only scant encouragement for his efforts. Undeterred, he continued his scientific studies.

Over time, Volta's youthful passion mellowed somewhat and he turned from grand theories to making instruments to explore the mysteries of nature so that by middle age, he was very much an instrumentalist first and a theorist second. Theories, he had apparently learned, were risky propositions while devices capable of reliably consistent results were very much coming into their own throughout Europe's scientific community. As instruments became more reliable and standardized, the emphasis shifted from theory to experiment. One of Volta's favorite maxims was, "The language of experiment is more authoritative than any reasoning: facts can destroy our ratiocination—not vice versa."

By the time Galvani's paper reached him, Volta was well into middle age and firmly established as a professor of physics at Pavia with a solid reputation as a skilled instrument maker. His scientific instruments, favorites among the well-heeled amateurs and serious natural philosophers, were beautifully built, though not particularly original. His specialty, developed over years of painstaking labor, was producing incremental improvements on already existing devices.

Adding to Volta's reputation was the meticulous manner in which he presented his devices. With each piece of equipment, such as his electrophorus, a machine for generating an electrostatic charge, he included precise instructions for its use as well as a detailed list of experiments to be carried out. Still something of a self-promoter, Volta made certain his devices reached influential people and even traveled widely to give public demonstrations.

Although deeply suspicious of science conducted by mere physicians, at the urging of colleagues he successfully duplicated Galvani's experiment in 1792, but maintained his reservations. Yes, the frog's leg may have twitched, Volta conceded, but it certainly wasn't because of electrical fluid held in the muscle. Some other, far more likely, explanation must be the reason. So the controversy began.

Given the participants, it was an odd debate. Galvani, the anatomist, had ventured into physics, while Volta, the physicist, was crossing over into anatomy. Cultured Europe, in which science was very much *salonfähig*, quickly began to line up on both sides of the issue.

Volta, for his part, was a strict adherent to the scientific method. Using a methodology very much like Franklin's disassembling of a Leyden jar, he discovered that electrical fluid generated in the frog experiment was a product of the sum of its parts rather than a single piece. In a series of experiments, he systematically substituted various components of Galvani's original experiment and soon found the secret resided not in the frog, but in the two dissimilar metals.

According to Volta, there was no "electrical-fluid imbalance" in the frog's anatomy. The frog played a largely passive role as instrument to detect minute levels of electrical current generated by the two metals. What Galvani had done in his original accidental experiment was to bring two dissimilar metals—his scalpel and a metal plate—in close proximity with a conductor, presumably fluid or tissue from the frog. As electrons were lost from one metal via oxidation and picked up by the other, the frog's nerve reacted to the flow of electrons—acting as a very sensitive voltmeter.

Placing two coins of different metals on his own tongue, he felt the distinct tingle of an electrical charge. Very soon Volta began ranking combinations of metals to determine which pairings produced the greatest electrical charge, or what he took to calling "electromotive force," and found a combination of silver and zinc seemed to offer the best results.

Galvani's error had been a reasonable mistake; he simply looked for the most likely source of the "electrical fluid" and settled on the once-living tissue rather than the inanimate metals. "He [Volta], in short, attributes everything to the metals, nothing to the animal; I everything to the former, as far as imbalance alone is concerned," Galvani wrote, summing up the controversy.

Sadly, Galvani was driven even further off course when he decided to see if an atmospheric electrical charge could make the frogs' legs twitch. Securing multiple frog legs to an iron fence by brass hooks, he detected slight twitching.

Then, in 1794, historians believe he followed up with an anonymous pamphlet that claimed the jerking of a frog's leg had been observed without any metal nearby, through simply touching the sciatic nerve of one leg to the muscle of another. In retrospect, this can be seen as the desperate tactic of someone heavily invested in a theory, but unable to offer solid scientific proof. Why else publish the pamphlet anonymously?

However, Volta, heavily invested in his opposing position, was also having difficulty supporting the bimetallic theory. The problem he faced was in offering definitive proof, which meant duplicating Galvani's experiment sans frog. In theory, it was easily accomplished; just substitute some other material for the moist tissue of an unlucky creature. Nothing could be simpler, except for the fact that the frog also acted as a voltmeter that measured the flow of electricity from one metal to the other. The frog may have played no significant function in the production of electricity, but its role in detecting electrical output was absolutely essential. At the time there was no other device, save the frog, sensitive enough to react to such low levels of electricity.

What Volta built was an updated version of an electrometer (also called an electroscope), created by William Nicholson, the English chemist and science writer. A simple device, it consisted of two sheaths of metal that attracted each other when an electrical charge was run through them. Volta's innovation was in using straw, rather than Nicholson's metal sheaths. He then combined the new meter with a device he invented for measuring atmospheric electricity called the *condensatore*, which was capable of collecting an electrical charge. A very simple device, the *condensatore* consisted of a metallic disk sitting flush on a nonconductive surface, such as marble. If a metal disk charged from a weak electrical source while on the nonconductive surface was lifted from the marble, the accumulated charge was greater than the current that flowed into it. Volta called this combination the "micro-electroscope."

At around the same time, Volta ordered a series of books from a dealer in Leipzig, Germany, including Nicholson's *Journal of Natural Philosophy, Chemistry and the Arts*. In it, the British writer meticulously outlined the likely workings of the torpedo fish (*Torpedo nobiliana)* known since ancient times for producing an electrical charge. Nicholson described the fish as containing anywhere from 500 to 1,000 columns or disks of a substance that was electrified oppositely, and divided by thin layers of a laminate. In his work, Nicholson went so far as to offer the conjecture that the "torpedo actually operates like a machine . . ." and then listed the individual components. It was an important point. In describing how the torpedo fish may have worked—essentially reverse engineering the creature—Nicholson sketched a rough blueprint for the first true battery.

With papers endorsing and refuting the two differing theories circulating between supporters of both sides, Galvani in some of his last writings proffered the idea of two different types of electricity—animal electricity and common electricity. Not quite a complete concession, the good doctor's suggestion was closer to a compromise based on diplomacy and perhaps weariness with the debate rather than experiment or scientific fact.

Volta, still working hard to prove his bimetallic theory, offered up his own compromise. He would admit that animal electricity exists, but not in the way that Galvani described—accumulating in muscles, particularly in severed limbs and small pieces of muscle. Yes, the nerves may act as conduits for electric fluid, Volta allowed, but that fluid doesn't originate in the muscles.

The debate probably lasted longer than necessary. Volta married, somewhat belatedly, at age forty-nine, and quickly fathered three sons in three years. The invasion of Italy by France also slowed his research when the occupying army closed the university in Pavia.

Then, on March 20, 1800, two years after Galvani's death, Volta settled the matter once and for all. In a thousand-word description in French, the language of the European scientific community, he described in detail the construction for a working battery. The letter was sent to Sir Joseph Banks, the president of the Royal Society, with whom he had carried on a long-standing correspondence. Volta's instruments were in wide use among the Society's members and he had, in fact, received the Society's highest honor, the Copley Medal, a few years earlier. A second letter soon followed that included a 5,000-word account, with diagrams. In this second letter, Volta compared his new device to the structure of the torpedo fish, a comparison he would make less frequently as time went on, and the real value of the battery in scientific research began to emerge.

The battery itself was a thing of stunning simplicity. Volta outlined its structure in his original letter in fewer than 400 words. Later, on May 30, a journalist for the *London Morning Chronicle*, penning the first description ever published regarding Volta's invention, described it in less than 150 words.

> A number of pieces of zinc, each the size of a half crown, were prepared, and an equal number of pieces of card cut in the same form; a piece of zinc was then laid upon the table, and upon it a half crown; upon this was placed a piece of card moistened with water, upon the card was laid another piece of zinc, upon that another half crown . . . Then a wet card, and so alternately until forty pieces of each had been placed upon each other; a person then, having his hands well wetted, touched the piece of zinc at the bottom with one hand, and the half crown at the top with the other; he felt a strong shock, which was repeated as often as the contact was renewed.

What Volta described in his letters to Banks would eventually become known as the "voltaic pile," though in the latter part of the letter, he also outlined the construction for another apparatus, which he called the "crown of cups"—essentially a battery whose components were distributed between different containers either stacked on top of one another or arranged side by side. In writing to Banks, Volta portrayed the battery simply as "an apparatus . . . which should have an inexhaustible charge, a perpetual action or impulse of electric fluid."

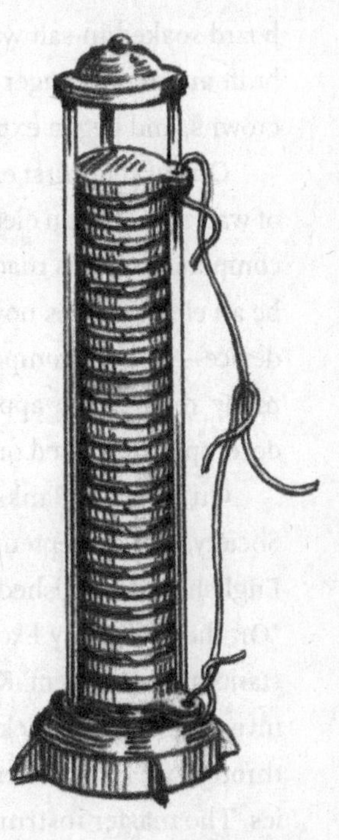

VOLTAIC PILE

Focusing nearly all of his attention on the instrument itself and its relevance to the controversy of animal versus bimetallic electricity, his theory of precisely how the instrument created the electrical charge was added almost as an afterthought. Volta's best guess was that the charge was generated through contact of two different metals. It was enough that the battery worked and that the debate was settled.

After receiving the initial letter in mid-April (the second would arrive several weeks later), Banks quickly began circulating it among friends and colleagues, including Anthony Carlisle, who was then a doctor at Westminster Hospital. From Carlisle, the letter made its way to Nicholson and on April 30, the two began constructing their own battery, per Volta's detailed instructions, made out of seventeen half crowns containing silver, several pieces of zinc, and plaster-board soaked in salt water. When this battery proved a success, they built an even stronger one on May 2, this time using thirty-six half crowns, and began experimenting.

One of their first experiments was the successful decomposition of water—using an electrical charge to break water down into its two components. This made big news across Europe. Water, thought to be an element, was now definitively shown—with the help of Volta's device—to be a compound composed of hydrogen and oxygen. In many reports the apparatus—the battery—that accomplished the decomposition rated only secondary mention.

On June 26, Banks read Volta's letter to the Fellows of the Royal Society, and by September of that year, the letter was translated into English and published in *Philosophical Transactions* under the title "On the Electricity Excited by the Mere Contact of Conducting Substances of Different Kinds." From there, news of the miraculous invention spread quickly and by the autumn of 1800 experimenters throughout Europe were building and using their own voltaic batteries. The master instrument maker had designed an instrument that nearly anyone could construct.

News of the batteries' success reached Volta not through British newspapers, but through a French paper called *Moniteur universel*, an official government publication. Volta, still true to his early instinct for self-promotion, set off on a tour of Europe to demonstrate his new device, traveling to Paris, London, and Vienna. He could not have picked a better time. In June 1800, Napoleon's army took Lombardy, and the Austrians, who had shuttered the University of Pavia, left the city. In late 1801, he demonstrated the battery for Napoleon himself, who had a keen interest in science, both personally and as a public relations tool. Napoleon's well-publicized interest in science seemed to have a somewhat calming effect on the ruling class, and in 1810, Napoleon made Volta a count.

Volta's public demonstrations, including an engagement at the

Royal Society, focused on the battery's design and function rather than theory and touched only lightly on the controversy over animal electricity that had lasted eight years. Volta carried with him a small, pocket-sized battery as well, showing how the device could be scaled down for portability—though its diminutive size proved less impressive than the larger pile or crown of cups.

Soon, other natural philosophers began giving their own demonstrations of the device. One of the strangest public performances was offered up by Étienne-Gaspard Robertson. An entertainer and amateur scientist showman, Robertson ran a popular show in Paris called *Fantasmagorie de Robertson*. The primary attraction was his use of "magic lanterns" mounted on dollies to simulate moving ghosts, though another feature included an electrostatic machine. After reading a description of the voltaic pile, he quickly commissioned one for himself, which he called *colonne métallique*, and began experimenting. Oddly, Robertson began his experiments by touching the battery contacts to various parts of his own body and those of volunteers, including chin, eyes, and ". . . other parts of the body where the skin is especially delicate and sensible."

Even as batteries were being built across Europe, there was still no clear consensus on just what to call the device. "New instruments should be given new names, depending not only on their form, but also on their effects or the principle on which they are based," Volta wrote, then suggested two possible names for his new device. His first choices were *organe electrique artificiel* as compared to the *organe electrique naturel* in reference to the torpedo fish. The second name he suggested was *appareil electro-moteur*. Neither caught on. A variety of names began cropping up, like "pile" or "trough," and then "voltaic pile," "galvanic battery," "Volta's battery," and other variations.

Volta never sought to patent his device. Indeed, the first patent for a battery would come several decades in the future, well into the nineteenth century and long after his death. Patents for electrical devices, of course, were difficult. It was necessary to show a useful application, and the few that did receive patents at the time often listed medical purposes.

NOT SURPRISINGLY, WITH A DEVICE capable of producing a continuous flow of electrical current, the debate surrounding animal electricity was soon forgotten. Unlike a Leyden jar, which required constant laborious charging and from which electricity exploded in an electrostatic burst, the battery was easily constructed out of readily available materials and provided a relatively long-lasting and steady flow of current with which to experiment. It's also interesting to note just how little desire there was to explore the way in which the battery produced its charge. Even the most serious experimenters focused almost solely on what that electrical charge could be made to do in the laboratory. The fact that it worked was enough.

With the Industrial Revolution in full coal-burning, steam-hissing swing, many saw the battery's future in medicine while others predicted electricity would eventually find use in movement and machinery with the potential to provide energy akin to steam.

However, the first practical applications for the battery were not in industry, but rather in science, specifically chemistry. Still not fully understood, the study of electricity was in its infancy and the battery found its first practical applications as a lab instrument. It allowed chemists to refine and redefine established theories, eventually laying the foundation of modern chemistry.

It would take decades before the battery would emerge in any significant role outside the lab, and even then, it came in an unpredicted area—that of communication via the telegraph.

# 4

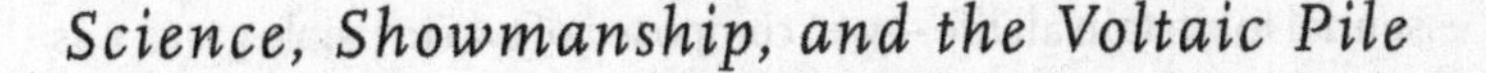

## *Science, Showmanship, and the Voltaic Pile*

*"More than the diamond Koh-i-noor, which glitters among their crown jewels, they prize the dull pebble which is wiser than a man, whose poles turn themselves to the poles of the world, and whose axis is parallel to the axis of the world. Now, their toys are steam and galvanism."*

—*Ralph Waldo Emerson,* English Traits

From the perspective of our technologically jaded age in which scientific and technical breakthroughs often rate little more than a perfunctory nod of acknowledgment, it is difficult to imagine the excitement science provoked among the general public during the early part of the nineteenth century. Scientific discoveries promised something far different from the engineering marvels of the Industrial Revolution that focused on such practical matters as increased production in automated mills or the speed and tonnage of locomotives.

Science was an exploration of the world beyond the borders of bare-knuckle nineteenth-century commerce and the routine concerns of everyday life. The promise proffered by nineteenth-century science was a greater and more intimate understanding of the world, specifically nature. It was not all that different from the adventurous scientific expeditions that fetched back specimens of insects, plants, animals, and artifacts from distant lands for examination and classification. And no science was more glamorous than chemistry.

Today schoolchildren learn that water is a combination of hydrogen and oxygen, yet for those at the start of the nineteenth century the discovery that the covering of three-quarters of the globe, upon which the great ships sailed to distant lands—and that the fluid essential to sustain all life—was composed of two invisible gases was a counterintuitive revelation.

The new study of galvanism was decidedly different from most branches of science. Though its precise nature was not yet fully understood, its potential was suddenly open to vivid, often far-ranging speculation and all manner of experimentation, thanks in large part to the voltaic pile and the Leyden jar. Periodicals and books intended for the nonscientist flourished, finding an eager readership among the middle class, which was already enthralled by the technological advances of the Industrial Revolution and the radical new philosophies sweeping through Europe and America. Crowds packed auditoriums to hear the latest theories on galvanism while newspapers reporting on the lectures ran stories with appropriate breathless awe. Memoirs written by those in London during the nineteenth century are packed with remembrances of lectures at the Royal Institution and news of the latest scientific breakthroughs.

Science was not only new and glamorous, but also held the potential to reveal secrets previously unknown to man—making it nearly blasphemous. As for the scientists, they made it a point to be as compelling as possible. The idea was to instruct through entertainment. Among the more spectacular of these science shows was that of Giovanni Aldini, the physicist nephew of Luigi Galvani. An early and ardent supporter of his uncle's theory of animal electricity, he surpassed experiments featuring twitching frogs, taking the same general principles to stunningly morbid levels. During one demonstration in the early 1800s Aldini applied current from a powerfully charged Leyden jar to the head of a freshly slaughtered ox, causing audiences to gasp as the eyes, nose, and tongue convulsed and spasmed.

Moving on to humans, he acquired from local authorities subjects fresh from execution for experimentation and demonstrations. "A large incision was made into the nape of the neck, close below the occiput," Aldini wrote of one early experiment on a recently deceased thirty-year-old man.

> The posterior half of the Atlas vertebra was then removed by forceps, when the spinal marrow was brought into view. A profuse flow of liquid blood gushed from the wound, inundating the floor. A considerable incision at the same time was made in the left hip through the great gutteal muscle so as to bring the sciatic nerve into sight, and a small cut was made in the heel; the pointed rod with one end connected to the battery was now placed in contact with the spinal marrow, while the other rod was placed in contact with the sciatic nerve. Every muscle of the body was immediately agitated with convulsive movements resembling violent shuddering from the cold . . . On moving the second rod from the hip to the heel, the knee being previously bent, the leg was thrown out with such violence as nearly to overturn one of the assistants, who attempted to prevent its extension.

In other experiments and demonstrations, Aldini reportedly used just the heads of executed prisoners. Moistening both ears with a brine solution, he completed a circuit with two wires—very much resembling twenty-first-century MP3 player headphones—attached to a crude battery comprised of a hundred layers of silver and zinc. "When this communication was established, I observed strong contractions in the muscles of the face, which were contorted in so irregular a manner that they exhibited the appearance of the most horrid grimaces," he wrote. "The action of the eye-lids was exceedingly striking, though less sensible in the human head than in that of an ox."

Traveling from Bologna to Paris, and then on to London, Aldini took his gruesome show on the road, staging scientific demonstrations featuring human cadavers and animals with theatrical flourish at universities and medical schools. Though usually not open to the general public, these demonstrations were excitedly reported in the popular press causing a sensation across Europe. In London, George Forster, found guilty of murdering his wife and child and duly executed, was rolled out onstage and into medical history at the Royal College of Surgeons fresh from the gallows at Newgate Prison. As Aldini began prodding the body with two rods attached to a charged Leyden jar, Forster's legs, mouth, and rectum clenched and contorted.

Aldini's efforts earned him the Royal Society's Copley Medal; to his credit, he never actually claimed to reanimate the dead, restricting his comments to more scientific phrases such as "command the vital powers" and "exerted a considerable power over the nervous and muscular systems." The *Times* of London also took a conservative view after witnessing his experiment in 1803.

> Its object was to shew [*sic*] the excitability of the human frame, when this animal electricity is duly applied. In cases of drowning or suffocation, it promises to be of the utmost use, by reviving the action of the lungs, and thereby rekindling the expiring spark of vitality. In cases of apoplexy, or disorders of the head, it offers also most encouraging prospects for the benefit of mankind. The Pro-

> fessor, we understand, has made use of *galvanism* also in several cases of insanity, and with complete success. It is the opinion of the first medical men, that this discovery, if rightly managed and duly prosecuted, cannot fail to be of unforeseen utility.

Yet the conclusions drawn by those outside the medical and scientific communities were often not as nuanced. If movement was tantamount to life, then life had been somewhat restored through electric shocks. Could it be possible that death was no longer permanent? Given the already uneasy relationship with the mystery of death prevalent during the nineteenth century, it seemed wholly possible that electricity might one day provide a viable alternative to the inevitability of the graveyard.

The Royal Humane Society, which had set up receiving stations and launched boats to assist drowning victims, added electric shock to its tools along with bellows and an early version of CPR. When Percy Bysshe Shelley's first wife, Harriet Westbrook, committed suicide by throwing herself into London's Serpentine, her body was reported to have been subjected to electric shocks from a voltaic pile in an attempt to revive her.

Hucksters of electrotherapy cures, plentiful in Europe and America, began plying their dubious trade with abandon, promising everything from a cure for impotence to arthritis. Even those with no scientific training or access to equipment recognized the public's gullibility when it came to electricity. One enterprising entrepreneur in London offered the services of a bathtub filled with torpedo fish for two shillings and sixpence.

As far back as 1759 John Wesley, the abolitionist and founder of the Methodist Church, wrote a book called *The Desideratum*, praising the ability of electricity to cure or alleviate diseases. An instant success, the book sold through five editions by 1781. Embracing electricity, which he called the "subtle fluid," as a miracle cure, Wesley shipped four electrostatic machines to London to treat angina pectoris, bruising, cold feet, gout, gravel in the kidneys, headaches, hysterics and memory loss, pain in the toe, sciatica, pleuritic pain, stomach pain, palpitations, and other common ailments. He concludes *The Desideratum*: "Let him for two or three Weeks (at least) try it himself in the above-named Disorders. And then his own Senses will shew him, whether it is a mere Plaything, or the noblest Medicine yet known in the World."

To many of its practitioners and observers, science was nearly a religious calling reserved for a fortunate and chosen few, and in early nineteenth-century London none was more select or lucky than the chemist Humphry Davy. Young, brilliant, and socially ambitious, as a protégé of the well-known physicist Joseph Priestley, he had invented a lamp that reduced the danger of explosions in coal mines. This was no trivial matter, since the Industrial Revolution was largely coal powered and explosions were a constant danger that claimed lives and slowed down production. Davy had also discovered the more pleasurable effects of nitrous oxide, which added to his scientific reputation as well as winning the friendship of literary luminaries such as Robert Southey and Samuel Taylor Coleridge.

Davy came from a family split between gentry and laborers. His father, a failed farmer, died when Davy was still in his teens, and he was handed off to a rich uncle, then handed off again, this time to a local surgeon and apothecary as an apprentice. Through a combination of good fortune and ingenuity, he ended up in exactly the right place at the right time.

Abandoning his provincial roots in Cornwall for London in the summer of 1800 with his reputation already secure, Davy was more than ready to begin a new chapter in science. Experimenting with Volta's device at the newly formed Royal Institution, he was appointed as chemist and assistant lecturer in 1801. In 1802, he was promoted to the position of professor, making him one of the first full-time, salaried, natural philosophers.

When first conceived, the Royal Institution of Great Britain was to be a showplace to exhibit and demonstrate commercial processes, in effect an open forum for showcasing the technological progress of the Industrial Revolution. This was not an altogether unreasonable proposition. Science had been contributing to industrial ventures in subtle and not so subtle ways since the 1700s. To cite a few examples, the simple acts of accurately measuring weight and temperatures proved a large benefit to a large number of businesses. However, with patent laws still not firmly established or even enforced, the idea of an open forum posed a problem. Manufacturing concerns, though benefiting from the technological advances of the day, still jealously guarded their processes as trade secrets; and almost immediately the Royal Institution drifted toward pure science with the managers allocating enough money to build one of the best-equipped labs in Europe.

Davy's lectures at the Royal Institution made him famous and his good looks made him something of a pop culture icon. "Those eyes were made for something besides poring over crucibles," women were reported to have said about him. As much as a serious scientist can be, he also seemed born for the stage. Neither a disheveled eccentric who labored alone and then stammered out his ideas nor a dusty, doddering, and droning professor, he offered audiences that most rare of all scientific presentations—passionate eloquence paired with impeccable showmanship.

Well mannered and dressed as fashionably as the audiences who flocked to see his demonstrations, he was a far cry from the secretive and unkempt Newton. Every inch a gentleman practicing what was still very much a gentleman's avocation, Davy took measures to ensure that only those worthy of his treasured ideas sat in attendance. He began charging admission to his lectures, then sealed a side door and tore down the balcony, both constructed expressly for the less fortunate classes.

Taking the stage before a packed audience, he used a voltaic pile to create sparks and set off small charges of gunpowder. Arcing current through two charcoal electrodes to produce a brilliant white glow was a favorite among audiences, even if they didn't understand that they were seeing the first electric light. The press covered his demonstrations in fawning detail while the scientific principles he explained, often in poetic terms (he was not above using the word "sublime"), became the focus of drawing room conversation.

So popular were the demonstrations that scalpers took to selling tickets at a hefty profit. On lecture nights at the Royal Institution, Albemarle Street suffered a nineteenth-century version of horse-and-carriage gridlock, prompting authorities to make it the first one-way street in London in an attempt to clear the congestion.

Davy was not only a talented showman of science, but also a brilliant experimentalist. Drawn to Volta's device, he quickly began experimenting with an improved design that incorporated alkaline and acid substances with a single metal plate. Napoleon, who had been so impressed with Volta's work that he established a prize through the French Institute, bestowed on Davy the medal along with a 300-franc prize—despite the fact that England and France were at war.

Using the Royal Institution's battery, Davy began decomposing compounds down into their constituent parts by way of electrolysis, which meant applying an electrical charge to the compounds. This eventually led to the discovery of five new elements and the isolation of other elements, such as lithium. This was groundbreaking, exciting stuff for the era. When an experiment in electrolysis on caustic potash and soda, which had resisted analysis, proved successful, Davy was more than a little overjoyed. His assistant later wrote, "When he saw the minute globules of potassium burst through the crust of potash and take fire . . . he could not contain his joy—he actually danced about the room in ecstatic delight: some little time was required for him to compose himself to continue the experiment."

It was these kinds of experiments that not only established Davy, but also seriously challenged the designation of chemistry as a

"French science" that had stood since Antoine-Laurent Lavoisier first drew the distinction between elements and compounds and then began the ordering of known elements in the 1780s. Ironically, Benjamin Thompson, the American-born British Loyalist who founded the Royal Institution, would later marry Lavoisier's widow after the chemist was pulled from his laboratory in the Louvre and beheaded during the French Revolution. "The Republic has no need of scientists," the judge was reputed to have said.

Davy's experiments also established that electricity and chemical affinity are identical. Although not the first to hit on this notion—again, the credit goes to Lavoisier—Davy was the first to prove it. In 1806 he demonstrated that decomposed water offers only two products—oxygen and hydrogen—in the same proportions as when water is synthesized. The conclusion was simple: electrical attraction holds elements together in compounds. This was a major step forward in understanding the nature of electricity as a vital force. Davy also expanded and refined Volta's view that electricity was generated by simple contact between differing metals. For Davy, the dedicated chemist, there had to be a chemical reaction at work.

THE ROYAL INSTITUTION—LIKE MANY organizations and individuals—began building larger and larger batteries. In July of 1808, only eight weeks after a hefty 600-plate battery was finished, Davy requested an even bigger voltaic pile, stating in his proposal that the "increase in size of the apparatus is absolutely necessary."

As passionate and tireless a fund-raiser as he was a lecturer and experimentalist, Davy played on his benefactors' pride in England's empire, comparing science to the great voyagers of exploration and a "country unexplored, but noble and fertile in aspect, a land of promise in philosophy . . . so rich a philosophy, and the useful arts connected with them."

In the age of Napoleonic conquest, he was also not above leveraging nationalistic fervor to raise the funds needed, rejecting all subtlety in calling the battery a critical piece of weaponry in the electrochemical and scientific war with France while reporting that Napoleon himself had ordered the construction of numerous large batteries at the École Polytechnique. "The scientific glory of a country may be considered in some measure, as an indication of its innate strength," he wrote. "There is one spirit of enterprise, vigour, and conquest in science, arts and arms . . . the same dignified feeling, which urges men to endeavor to gain dominion over nature will preserve them from humiliation and slavery."

It is worth noting that scientific ideas seemed to be flowing with relative freedom across borders, even between nations engaged in hostilities. Carried by smugglers and business concerns, periodicals and letters detailing the latest scientific discoveries moved freely during the Napoleonic Wars from 1803 to 1815, in some cases traveling more rapidly between England and France than from Switzerland.

Davy's request for a larger battery was approved and in December of 1809 he was presented with a battery of some 2,000 six-inch-square double plates. At the time it was the world's largest battery, built by donations to a subscription fund at the Royal Institution.

One of a new breed of natural philosophers, Davy recognized the impact that science could have on industry and society. In addition to redesigning the miner's lamp, he had studied the chemical processes of tanneries and agriculture, published the *Elements of Agricultural Chemistry* in 1813, and even tried to solve the problem of shipworms, a major cause of concern in the shipping industry, by attaching positively charged iron and zinc to the copper hulls of boats.

Also, like many others of his era, Davy saw no impenetrable line of demarcation between science and the arts. Rather, he and others of his time saw them as components of a whole.

> Man, in what is called a state of nature, is a creature of almost pure sensation. Called into activity only by positive wants, his life is passed either in satisfying the cravings of the common appetites,

> or in apathy, or in slumber. Living only in moments he calculates but little on futurity. He has no vivid feelings of hope, or thoughts of permanent and powerful action. And unable to discover causes, he is either harassed by superstitious dreams, or quietly and passively submissive to the mercy of nature and the elements. How different is man informed through the beneficence of the Deity, by science and the arts!

Gentlemen of the "better classes" were still expected to possess a working knowledge of science, along with art, history, and literature. It was an era in which ideas really mattered, and understanding the mysteries of nature through scientific exploration was a very big idea. Conversely, many artists, writers, and poets who had traditionally focused their attention and creative powers on myths turned their talents to depictions of science while those laboring away in the lab dabbled in the arts. Davy wrote poetry. In one poem, he pays tribute to nature, which was slowly yielding its secrets through scientific exploration.

> *Oh, most magnificent and noble Nature!*
> *Have I not worshipped thee with such a love*
> *As never mortal man before displayed?*
> *Adored thee in thy majesty of visible creation,*
> *And searched into thy hidden and mysterious ways*
> *As Poet, as Philosopher, as Sage?*

A friend of William Wordsworth, Davy persuaded the poet that natural philosophers were akin to poets in their search for meaning, and later edited Wordsworth's *Lyrical Ballads,* while Coleridge attended scientific lectures in search of metaphors. And in America, Ralph Waldo Emerson, starting his first journal after leaving Harvard in 1820, listed Davy's *Elements of Chemical Philosophy* as a book he intended to read.

Percy Bysshe Shelley, perhaps the most romantic of the Romantic poets, was himself an enthusiastic amateur natural philosopher for much of his short life. As a young boy he enlisted his sister, Helen, and her playmates in experiments with a charged Leyden jar. Later she would write, "[My heart] would sink with fear at his approach; but shame kept me silent, as with many others as we could collect, we were placed hand-in-hand around the nursery table to be electrified." His love of science continued through his days at Eton and then Oxford, where his rooms were crowded with microscopes and pumps.

Later, his wife, Mary Wollstonecraft Shelley, who would write *Frankenstein, or, The Modern Prometheus,* took an equally ardent interest in science. Though often portrayed as an innocent because of her youth when she married the poet, Mary Shelley was, in fact, far from naïve. (She began writing *Frankenstein* when she was eighteen, and it was published when she was twenty-one.) She would have more than held her own with her husband, along with Lord Byron, as well as the somewhat sinister doctor John Polidori during that rainy 1816 vacation at the Villa Diodati on Lake Geneva where the idea of Frankenstein took hold during a parlor game of ghost stories.

Given her upbringing, Mary Shelley would have been quite at home with such company. Her father, William Godwin, a former minister turned ardent atheist, advocated a particularly idealistic philosophy inspired by the "scientific principles" of the French Revolution. Based on reason, justice, and universal education, it called for the peaceful overthrow of all religious, political, and social institutions. Coleridge, along with the writer Mary Lamb and the former American vice president Aaron Burr were frequent visitors to her father's house, as were Humphry Davy and William Nicholson.

If Mary Shelley had not actually witnessed Aldini's gruesome public exhibitions, she was certainly aware of them and was familiar with the scientific principles of Davy's work. Although she is coy about the exact process used to animate the monster, as she noted in her diary during the writing of *Frankenstein*, she was reading Davy's *Elements of Chemical Philosophy*.

"Before this I was not unacquainted with the more obvious laws of electricity. On this occasion a man of great research in natural philosophy was with us, and excited by this catastrophe, he entered on the explanation of a theory which he had formed on the subject of electricity and galvanism, which was at once new and astonishing to me," she has her protagonist write.

In a subsequent 1831 edition of her book, she wrote in the introduction, "Perhaps a corpse would be re-animated; galvanism had given token of such things: perhaps the component parts of a creature might be manufactured, brought together, and endured with vital warmth."

Edgar Allan Poe was less coy about the methodology for bringing the deceased back to life by way of an electrical charge. In his story "Some Words with a Mummy" he becomes nearly clinical on the methods used to bring a mummy back to life.

> Hereupon it was agreed to postpone the internal examination until the next evening; and we were about to separate for the present, when some one suggested an experiment or two with the voltaic pile. The application of electricity to a mummy three or four thousand years old at the least, was an idea, if not very sage, still sufficiently original, and we all caught it at once. About one-tenth in earnest and nine-tenths in jest, we arranged a battery in the Doctor's study, and conveyed thither the Egyptian . . . It was only after much trouble that we succeeded in laying bare some portions of the temporal muscle which appeared of less stony rigidity than other parts of the frame, but which, as we had anticipated, of course, gave no indication of galvanic susceptibility when brought in contact with the wire.

Likewise, Lord Byron, who reached a nineteenth-century version of rock star fame at the height of his career—"Mad, bad and dangerous to know" according to Lady Caroline Lamb—was also keenly aware of scientific advances, though he seemed to view them with more than mild suspicion. In his masterpiece *Don Juan*, he wrote,

*Bread has been made (indifferent) from potatoes;*
*And galvanism has set some corpses grinning,*
*But has not answer'd like the apparatus*
*Of the Humane Society's beginning*
*By which men are unsuffocated gratis:*
*What wondrous new machines have late been spinning!*
*I said the small-pox has gone out of late;*
*Perhaps it may be follow'd by the great*

And then, a few stanzas later,

*This is the patent-age of new inventions*
*For killing bodies, and for saving souls,*
*All propagated with the best intentions;*
*Sir Humphry Davy's lantern, by which coals*
*Are safely mined for in the mode he mentions,*
*Tombuctoo travels, voyages to the Poles,*
*Are ways to benefit mankind, as true,*
*Perhaps, as shooting them at Waterloo.*

And, for Herman Melville, electricity—mysterious and powerful as it seemed at the time—served as a perfect metaphor for Captain Ahab's primal obsession and madness, which he transmits through the crew as if through an electrical circuit in *Moby-Dick*.

"Advance, ye mates! Cross your lances full before me. Well done! Let me touch the axis." So saying, with extended arm, he grasped

> the three level, radiating lances at their crossed centre; while so doing, suddenly and nervously twitched them; meanwhile, glancing intently from Starbuck to Stubb; from Stubb to Flask. It seemed as though, by some nameless, interior volition, he would fain have shocked into them the same fiery emotion accumulated within the Leyden jar of his own magnetic life. The three mates quailed before his strong, sustained, and mystic aspect. Stubb and Flask looked sideways from him; the honest eye of Starbuck fell downright.
>
> "In vain!" cried Ahab; "but, maybe, 'tis well. For did ye three but once take the full-forced shock, then mine own electric thing that had perhaps expired from out me. Perchance, too, it would have dropped ye dead . . ."

Electricity was still a phenomenon to be studied without much thought as to its eventual value in the marketplace. Researchers advanced differing views and theories. Was it a chemical or mechanical force? Mysterious and invisible, electricity still seemed to have little apparent use. Volta had ignited gunpowder and then severed a thin length of wire with a charge from his voltaic pile during a demonstration for Napoleon, but these were mainly parlor tricks that only hinted at some future use.

In Italy, Luigi Brugnatelli, a professor of chemistry at Pavia and a friend of Volta's, managed to develop a form of electrodeposition (electroplating) just a few years after the development of the voltaic pile. The metallic object to be coated—say a trophy—was put into a solution filled with dissolved salt of the metal to coat it. The trophy was then negatively charged by way of a battery while the positively charged pole made up of the plating material—for example, nickel or silver—gave up electrons. Since opposites attract, the positively charged silver or nickel coated the object.

Brugnatelli's idea was ingenious. Unfortunately, his findings were repressed because of a dispute with the French Academy of

Sciences. It would fall to British and Russian scientists to come up with the principles independently years later.

THE BATTERY, SO SIMPLE IN design, quickly developed into a field of research in itself as natural philosophers sought more power and longer-lasting batteries. The key, the researchers knew, was in combining different types of metals and solutions. In 1802, Volta discovered that manganese dioxide and zinc in a saline solution generated a higher voltage than just copper and zinc. Within a remarkably short period of time, improved battery designs began emerging from labs across Europe.

At around the same time, William Cruickshank, a Scottish chemist and military surgeon, read Volta's letter and set about improving the original design. Building a grooved wooden box, he soldered copper and zinc plates horizontally, and then flooded the box with an acidic solution. Although imperfect—the box tended to leak at the seams—it did provide stronger current than Volta's original design and proved to be the first battery capable of something like mass production by instrument makers.

William Hyde Wollaston was a physician who had given up his practice in favor of pure science in the fields of chemistry, physics, and physiology. He discovered platinum and palladium early on, and his work in crystalline structures yielded not only scientific breakthroughs, but also instruments for measuring crystals. Turning his attention to lenses led him to create the camera lucida (Latin for light), a tool that allowed artists to draw in more accurate proportion.

The battery he developed in mid-1815 or 1816, known as the Wollaston pile, featured copper plates bent in two with the zinc placed between the two halves like a sandwich, though the metals were kept from touching by way of a wooden dowel. Like Cruickshank's battery, the whole arrangement was mounted in a wooden trough and submerged in an acidic solution.

In 1836, John F. Daniell, a professor at King's College London, came up with a cell that produced a steadier, longer-lasting current than Volta or Cruickshank's devices. Until then, battery chemistry had posed an inherent and mysterious problem. Hydrogen released from the chemical reaction would accumulate on the "passive" copper plate. These bubbles would eventually build up and block the current. Solving the problem was easy but usually meant removing the plate (or plates) and wiping it down on a regular basis.

Daniell's solution was an entirely new battery design that consisted of a cylindrical copper vessel with a porous earthenware container inside that held a length of zinc rod. He then filled the space between the copper and the porous vessel with a solution of copper sulfate saturated by salt crystals placed on a perforated shelf. The porous cup was filled with dilute sulfuric acid. It was a brilliant, though somewhat overly complex design. The earthenware kept the fluids separated and the current flowing. Called the "constant battery," the device had to be taken apart when not in use in order to halt the chemical reaction.

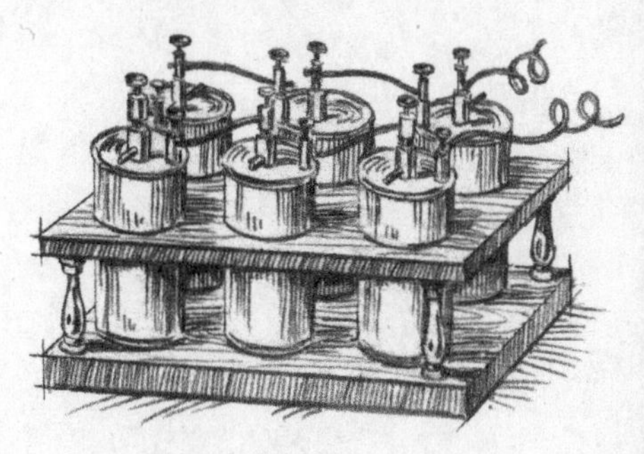

DANIELL BATTERY

SIR WILLIAM ROBERT GROVE, A lawyer and judge with an interest in science, also tried his hand at battery design in the 1830s. His first cell was a zinc and platinum combination with the zinc in sulfuric acid and the platinum in nitric acid, divided by a Daniell-like porous container. Although creating a powerful charge of about 1.8 volts, the chemical reaction tended to release poisonous nitric dioxide gas. Still, the battery would eventually become a favorite among telegraph companies.

Later, in 1839, Grove would invent what is commonly known as the first "fuel cell." It was already an established fact that water is split into its components of hydrogen and oxygen by an electrical current. Grove reversed the operation by combining oxygen and hydrogen to produce electricity and water. His design was brilliant. He fitted two platinum strips in a closed tube filled with hydrogen and oxygen in sulfuric acid. It worked, though not well enough for commercial production.

Daniell as well as Grove saw only limited use for their new and very much improved batteries. However, that was about to change in a dramatic way.

# 5

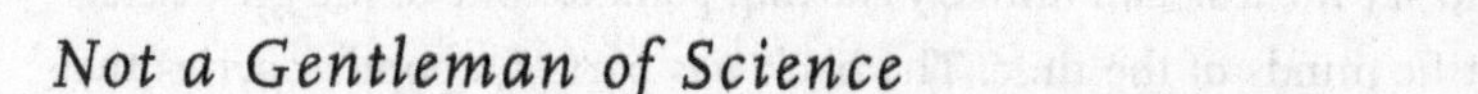

# *Not a Gentleman of Science*

*"Work, finish, publish."*

—*Michael Faraday*

Arguably, one of Davy's greatest and most unlikely discoveries was Michael Faraday. Nothing in Faraday's early life suggested a career in science or the slightest similarity to the well-bred gentlemen who dominated natural philosophy in England. He had neither a comfortable family fortune nor a degree from Cambridge or Oxford. The son of a blacksmith, his formal education had ended at thirteen with his entrance into an apprenticeship to George Ribeau, a Blandford Street bookseller in London in 1804.

Those were the days when bookstore patrons chose the bindings of their books according to taste and budget, and for eight years the young man was kept busy at the glue pots and presses. He could read and do basic sums, necessities in the bookbinding trade, but not much more. However, the young Faraday possessed a quick mind—recognized very early on by his employer—and a nearly insatiable curiosity. What little free time was available to him was spent reading the books in the shop, which set him off on a lifelong quest of self-improvement. He began keeping a journal after it was suggested in a book called *The Improvement of the Mind*, but it was science that really captured his imagination.

"There were two [books] that especially helped me," Faraday would write much later in life. "*The Encyclopaedia Britannica* from which I gained my first notions of electricity, and Mrs. Marcet's *Conversations on Chemistry*, which gave me my foundation in that science."

This was the kind of admission that must have at least raised an eyebrow at the time. Jane Marcet's *Conversations on Chemistry* was, to say the least, an unlikely starting point for one of the great scientific minds of the time. The book was part of a popular series that included *Conversations on Political Economy*, *Conversations on Vegetable Physiology*, and *Conversations on the History of England*.

*Conversations on Chemistry* was a far cry from the somber lecture halls and even more somber dons of Oxford or Cambridge. Written "Particularly for the Female Sex," it was a book by a woman for women. Although there wasn't much chance of offending delicate female sensibilities, even by early nineteenth-century standards of public morality, with talk of compounds and elements, the book assumed little or no prior knowledge of chemistry. Written in a conversational style as a series of genteel conversations between the kindly instructor, Mrs. Bryan (called "Mrs. B."), and her two eager and ever curious pupils, Emily and Caroline, readers were patiently walked through the very basics, step by well-mannered step. An international bestseller of the day, the book went through an enviable sixteen editions with more than 160,000 copies sold in the United States alone.

Science, particularly chemistry, was still a popular topic of dinner and drawing room discussion. The fictional Mrs. B. and her pupils educated legions of women (and probably a few men) well enough to hold up their end of the conversation. Well-bred men, presumably, did not need feminine primers for their education in such serious matters, having received the basics at college that allowed them to keep up with the latest developments in a suitably masculine fashion.

Access to books was only one piece of Faraday's early luck. By all accounts, Ribeau was an ideal employer. Unmarried and childless, he fully intended to eventually turn over the business to the bright young man, but Faraday's mind was firmly set on a career in science. That he was willing to risk the security of an established business—a very good proposition for the son of a blacksmith—for the uncertainty of a career in science was more than just a young man's optimism.

He was that most rare of creatures, a true believer when it came to science and religion. A member of a gentle Christian sect, the Sandemanians, Faraday was deeply religious and viewed science—the exploration of nature—as an extension of his heartfelt faith. Although we in the twenty-first century debate the conflict of science and religion, Faraday saw no such division. "The book of nature, which we have to read, is written by the finger of God," he wrote. For Faraday, "unraveling the mysteries of nature was to discover the manifestations of God."

Filled with the kind of youthful pluck found in the American Horatio Alger books—though seldom in real life—Faraday wrote to Joseph Banks, then president of the Royal Society, requesting a job, any job. When no response was forthcoming, he went to the Royal Society in person and was told by a functionary that Mr. Banks had said "the letter required no answer." Apparently, the young bookstore apprentice was beneath consideration, unworthy of even a reply.

Then came one of those seemingly small events upon which lives turn. One of the bookstore's customers, William Dance, gave the young man tickets to Davy's last series of lectures in March and April 1812. Faraday not only eagerly attended, but also took copious notes. Practically transcribing each lecture word for word, he returned to the bookstore and quickly bound the notes—nearly 400 pages complete with misspellings and his own illustrations—and sent the book to Davy along with another letter asking for a job.

Like a character out of a Dickens novel whose life is lifted on a fulcrum of unlikely chance, Faraday was in the right place at precisely the right time. One of the Royal Institution's lab assistants had recently gotten into an unseemly brawl with an instrument maker and was promptly fired. A replacement was needed and Faraday was installed in the lab as a "scrub" to carry out the drudgework of washing and sorting bottles.

Davy, like the kindly bookstore owner, quickly recognized Faraday's potential and was soon giving the young man analyses to make on his own, guiding him through experiments. The dexterity and precision learned at bookbinding proved good training for the budding experimentalist, and more experiments followed. Davy, like Ribeau, was childless and saw in Faraday a talented protégé.

As for Faraday, he idolized Davy, but their relationship was often uneasy. In one of the stranger episodes, the newly married Davy, at the end of his tenure at the Royal Institution, took Faraday along as an assistant on a trip abroad in 1813. Receiving special permission from Napoleon, Davy and Faraday stopped in France on the way to Italy. The trip, by some accounts, was a horror for the unworldly Faraday, who had never been far from London. His hero, Davy, was apparently cordial enough, but the new Mrs. Davy treated her husband's protégé more as a servant than a lab assistant. In letters he wrote home, Faraday bitterly bemoans her as "haughty and proud to an excessive degree." Returning to England, Davy was installed as president of the Royal Society, a post that Newton had once held, though not before securing Faraday the superintendent's position at the Royal Institution.

Tireless and always curious, Faraday worked vigorously in the lab, adopting the motto, "Work, finish, publish." Even by the standards of the day, which prized hard work and results, his reputation for long hours of experimentation was quickly established. "Let no one start at the difficult task and think the means far beyond him," Faraday wrote. "Everything may be gained by energy and perseverance."

Although lacking the formal ambition of Davy, and refusing most honors and accolades, Faraday kept up a lifelong routine of self-improvement. When he began lecturing at the Royal Institution, he placed his elocution teacher in the front row to critique the performance later. Friends were also invited and held up cards during the lectures printed with the words "Too fast" or "Too slow." His lectures eventually surpassed even those of Davy in popularity, attracting the attention of Charles Dickens, who would later request his notes to publish in *Household Words,* a general-interest magazine he published for nine years. More intriguing still was a brief flirtation Faraday carried on with the brilliant Augusta Ada Byron Lovelace, Lord Byron's daughter, who was a pioneer in computing theory.

Even as his fame and reputation grew, Faraday's relationship with Davy remained difficult in ways that went beyond the usual mentor and protégé. No matter how brilliant his lectures and experiments, in Britain's rigid social order Faraday was decidedly not one of the gentlemen of science. This point was made clear when Davy and William Hyde Wollaston tried an experiment and failed. Faraday, overhearing the details, went on to complete the experiment successfully, then committed the grave error of writing up the paper without crediting his social betters. Making the situation even worse was the fact that it was no ordinary experiment, but a breakthrough.

Faraday had apparently heard Davy and Wollaston discussing the work of the Danish medical professor Hans Christian Ørsted. During a lecture to his students in 1820, Ørsted left a compass near a completed circuit running from a voltaic pile and noticed that the compass needle twitched when moved closer to the circuit.

It was a major discovery. Electricity, it seemed, was not like water in a sealed pipe; fully contained within the confines of a conductor.

Rather, it created an invisible field of emanating waves around whatever carried it. What Ørsted discovered was the electromagnetic field and electromagnetism. From this phenomenon, all manner of devices were possible.

After hearing about the discovery, the French scientist André-Marie Ampère created a magnetic field by forming wires into coils. He later invented the galvanometer—a voltmeter—by simply taking a few turns of wire around a compass. The slightest amount of current would create an electromagnetic field and cause the needle to twitch. This was a major technological step forward in measuring minute currents. A few crude voltmeters and other methods of measuring substantial currents (such as how quickly they melted a length of thin wire) did exist, but detecting very small charges presented a problem. Over the years scientists had taken to applying live wires to the tongue and tasting the voltage or even judging the pain it caused by inserting wires into tiny cuts in the skin. Ampère's invention was the most sensitive measurement device of electric current since Galvani's frog legs.

In fact, Ørsted was not the first to notice the phenomenon. In 1802 the Italian jurist Gian Domenico Romagnosi had made a very similar discovery, but he published his findings in an obscure journal, *Gazetta di Trentino,* and they went largely unread by others working in the field. Ørsted's independent discovery and subsequent publication would change not only the way small currents were measured but also would lead the investigation of electricity off into new directions. Although the voltaic cell was well established as a scientific instrument, it was still the chemists who were reaping the majority of rewards from its use. The idea that electricity could prove a practical force beyond the study of the molecular level of compounds and elements did not occur to the majority of those working in the field.

In the mid-1820s, William Sturgeon, a shoemaker's apprentice who ran away to join the Royal Army, invented the electromagnet based on reading about Ørsted's discovery. The self-taught Sturgeon, who was by then lecturing at the Royal Military College, wrapped a few turns of wire attached to a battery around a seven-ounce piece of iron and lifted nine pounds. Electricity—invisible, weightless, and mysterious—could be made to perform labor.

The concept Davy and Wollaston discussed, but Faraday proved, was to make a wire rotate within a magnetic field, in effect creating a simple electric motor. Faraday placed a magnet upright in a cup of mercury connected to one terminal of a battery. The second terminal was connected to a wire with one end in the mercury. When the circuit running through the wire, mercury, and battery was completed, the wire began rotating around the magnet. In another version he made a magnet rotate around a wire.

Faraday called his gadget a "rotator" (today called a homopolar motor) and the paper he wrote detailing its function brought him international fame. Later, when William Gladstone, the future prime minister, asked what possible good his tiny electrical motor was, Faraday was said to have responded, "I have no idea, but no doubt you'll find some way to tax it."

Wollaston's fury at Faraday's success went far beyond the natural rivalry still common among practitioners of science today. Faraday had clearly defied the class system so much a part of Regency London. What's more, Faraday lacked not only a gentleman's education from Oxford or Cambridge, but also the knowledge of advanced mathematics that came with it—including Newton's inspired achievement, calculus. For many in the Royal Institution, Faraday was clearly not one of the gentlemen of science, but rather, a talented upstart tinkerer.

Wollaston, by contrast, was not only of the correct background, but was also brilliant with a long string of discoveries to his name. Realizing his mistake, Faraday wrote an anguished letter of apology, but the damage was done. The letter elicited only an icy response and Davy, siding with Wollaston, twice blocked Faraday's membership into the Royal Society.

Much later, when preparing a book of his work, Faraday would add a footnote to the paper stating, "This is a very precious paper for me. I published it as a result of work given to me by Humphry Davy at a time when my fear far exceeded my knowledge." At the time, both Davy and Wollaston were long dead and the feud largely forgotten by all but Faraday.

Undeterred, Faraday kept hard at work and in August 1831, he hit upon the idea of induction, proving the principle behind a generator creating an electrical current by moving a magnet inside a coil of wire. The idea was simple—if electricity could produce magnetism, as Sturgeon had clearly demonstrated—then magnetism should produce electricity. He took a paper cylinder and wound it with coils of wire, then connected it to a battery and a primitive voltmeter. He then began moving a bar magnet in and out of the hollow center of the tube, making the needle of the voltmeter jump. Somehow the simple act of moving the magnet had created a burst of electrical current in the coil.

A year after Faraday announced his discovery, Hyppolyte Pixii, a young Parisian instrument maker, created the first electric generator by spinning a magnet over a coil with a hand crank. Within a decade, improvements on Pixii's design made industrial generators possible.

Faraday was also bringing a new scientific language into use with the help of William Whewell, the esteemed philosopher and teacher, who provided the words for battery components still in use today, including "anode" (the negative electrode that gives up the charge) and "cathode" (the positive electrode that accepts the electrons). This was no trivial matter. A standardized terminology is an essential component for scientific and technical research.

Before he ended his career in quiet retirement, Faraday had made enormous strides in the field of electrochemistry. Following in the footsteps of his mentor, Davy, he discovered the elements sodium, potassium, calcium, and magnesium. He also set down several laws regarding electrolysis—the separation of compounds into their elements—in *Faraday's Law of Induction and Laws of Electrolysis.*

"Electricity is often called wonderful, beautiful; but it is so only in common with the forces of nature," he wrote in his lecture notes. "The beauty of electricity, or of any other force, is not that the power is mysterious and unexpected, but that it is under law, and that the taught intellect can even now govern it largely. The human mind is placed above and not beneath it."

Sadly, in the end, Faraday was left behind. Mathematics was quickly becoming the language of science as investigations into nature became more and more complex, and Faraday just could not keep up. In a letter written toward the end of his life to the eminent scientist James Clerk Maxwell, he wrote,

> There is one thing I would be glad to ask you. When a mathematician engaged in investigation of physical actions and results has arrived at his own conclusions, may they not be expressed in common language, as fully, clearly and definitely as in mathematical formulae? If so, would it not be a great boon to such as we to express them so—translating them out of their hieroglyphics that we might work upon them by experiment?

By the 1830s, science was changing. Electricity, though still an inexplicable force, was offering hints that it could, very possibly, be useful for something. After all, it had been made to perform the simple physical labor of lifting things with an electromagnet. Science was also expanding beyond philosophy. Whewell, philosopher and friend of Coleridge and Faraday, coined the word to describe what was happening. The word was "scientist." It was not a word that caught on quickly among the gentlemen of science who viewed the study of the physical world as an extension of abstract reasoning and themselves as philosophers unraveling nature's innermost secrets. It had a tradesman's sound to it and was often used as a pejorative.

In America, Faraday's counterpart, Joseph Henry, was also hard at work. Although divided by an ocean and thousands of miles, Henry and Faraday led strangely parallel lives. Born within just a few years of each other—Henry in 1797 and Faraday in 1791—the two scientists saw their most productive years as well as their research overlap.

That Henry did not attain the same historic stature as Faraday does not diminish his contributions. Few scientists appear in history books alongside inventors such as Thomas Edison, Henry Ford, or Samuel F. B. Morse. Names like Albert Einstein, Stephen Hawking, and Isaac Newton are among the handful of exceptions that attest to the rule.

One reason is the basic fact that to a large degree, the most enduring legacy of science is knowledge. Scientific experimentation, abstractions, and discovery of underlying principles hold little popular appeal today compared to products that transform everyday life or create vast fortunes. Successful inventors leave behind foundations and museums while successive, evolving versions of their original devices carry their name forward. For better or worse, popular history belongs to the clever engineers who successfully apply scientific principles and not to the scientific explorers.

Although Henry, like Faraday and Davy, dabbled with inventions, his contributions were largely obscure or were overshadowed by other well-known names who perfected his concepts for the marketplace. While the media of the day in Europe was heaping praise on Faraday and his scientific demonstrations, Henry labored in relative obscurity for much of his most productive years. He was slow to publish, favoring the immediacy of teaching and experimentation. And, too, unlike Faraday, he was not at the center of Europe's tightly linked scientific establishment. Though not quite the scientific backwater it had been during Franklin's day in the eighteenth century, America still lagged behind Europe when it came to pure research.

Both Faraday and Henry were raised in less than affluent families with no tradition of education. Born in upstate New York, near Albany, Henry, like Faraday, lost his father and apprenticed at an early age. However, unlike Faraday, Henry's apprenticeship to a watchmaker and silversmith by the name of John F. Doty held little interest or opportunity for the lad. The watchmaker, in sharp contrast to Faraday's well-intentioned Mr. Ribeau, found the young man singularly unsuited for the work. While Faraday had taken an interest and exhibited genuine talent for the book business, Henry was notably less enthusiastic when it came to his trade. He tended to daydream at the workbench and showed no particular aptitude for the type of meticulous labor required in dealing with escapements and springs.

In any event, it didn't matter. Doty's business went under in the panic of 1819, terminating Henry's apprenticeship after just two years. Without a clear career path, the youth seemed to drift from job to job, working as a handyman, metalworker, and surveyor. An avid amateur actor, at one point he even toyed with dedicating himself to a life on the stage. Then, according to several slightly varying legends, he found his way to science, like Faraday, through a book. In one version of the story, he happened to climb down into a church basement to find the book lying about. In another account, a boarder in the house where he lived left the book behind. Whether it was through divine providence or simple carelessness, Henry was hooked. The title, according to Henry's own recollection, was *Lectures on Experimental Philosophy, Astronomy and Chemistry* by G. Gregory, D.D., Vicar of Westham.

In the same way that Ms. Marcet's primer inspired Faraday to take up science, the good vicar's tome had a similar effect on Henry. But while Faraday had the benefit of finding himself at the Royal Institution, Henry's newly discovered ambition was provided no such opportunity. Finally, after finishing his education at the Albany Academy for Boys, he drifted into teaching mathematics and natural philosophy at the school while supplementing his meager income by tutoring. Among his notable students was Henry James, the theological philosopher and father of William James, the philosopher, and Henry James, the writer.

Henry also began a series of experiments in chemistry and meteorology in his spare time, though not much seems to have come of these inquiries. Then, on a trip to New York City in 1827, he happened to see a demonstration of Sturgeon's electromagnet. Here was an ordinary piece of iron suddenly given life and power through a small voltaic pile—a simple chemical reaction. Fascinated, Henry managed to duplicate the wondrous device, voltaic pile and all, after reading about it in the journal *Annals of Philosophy*. Like Sturgeon's original, Henry's replica lifted nine pounds. It was the first electromagnet built in the United States.

Not content to simply copy the device, he began experimenting. The trick to increasing the magnet's strength, he discovered, was in the wire coil or helix that encased the horseshoe-shaped iron core. Unlike Sturgeon, who had taken only a few loose turns of wire around the core, Henry began wrapping the wire tightly against the iron and using more of it in the process. In the days prior to insulated wires he managed to avoid short circuits by varnishing the iron core and carefully keeping each turn of the wire separated. In the end, he found the device could be made to lift twenty pounds with the same power source that had lifted nine.

Encouraged by this small success, he kept going, replacing the type of primitive, low-output battery Sturgeon had used for a larger Cruickshank trough with twenty-five pairs of zinc and copper plates to provide more power, and winding the wire tighter and tighter around increasingly larger iron cores. When the wires began crackling with current, he insulated them (at least according to legend) with lengths of silk ribbon from his wife's petticoats. His experiments took him beyond anything Sturgeon, or even Faraday, had accomplished when it came to electromagnetism.

For instance, through trial and error he discovered that if a single cell powered the battery, it was best to use multiple lengths of wire wound tightly around the core in parallel. However, when using multiple cells, the electromagnet performed best when wound with a single long strand of wire.

Soon after, Henry came across what was to be one of his greatest discoveries. Creating a parallel circuit—multiple batteries attached positive to positive—the voltage remained the same no matter how many cells or batteries he wired, but the amperage (amps) increased in proportion to the number of connections made. Conversely, by creating a series circuit—the positive terminal connected to the negative terminal and the negative to the positive—he doubled the volts and got the same amperage. To use the common, though somewhat inaccurate, analogy: if an electrical circuit is like water in a pipe, then volts are a measure of the water pressure and amps a measure of the volume of water moving through the pipe. Henry called these difference types of measurements "intensity" and "quantity." The use of "volts" and "amps" would not become the official terminology until 1893.

In a very short time Henry and his assistant, Philip Ten Eyck, built an even larger magnet that weighed 21 pounds and was capable of hoisting some 750 pounds when mounted on a scaffold. For those who had never seen an electromagnet, which included just about everyone at the time, the device was very much like a magician's trick. Proof that the seemingly simple device powered by just a battery had actually lifted a blacksmith's iron anvil was provided when

Henry cut the power to the magnet and sent the anvil crashing dramatically to the floor.

This was not the fluttering of a compass needle or the playful attraction of a lodestone, but power on a scale to inspire awe. It is not difficult to imagine those who witnessed Henry's demonstration being as "thunderstruck" as St. Augustine when introduced to the diminutive power of a lodestone. Calling it the "Albany magnet," Henry wrote a paper detailing its design and sent it off to Benjamin Silliman, professor of chemistry at Yale University and editor of the respected *American Journal of Science*. In the editor's note accompanying the paper's publication in January of 1831, Silliman pointed out, "He [Henry] has the honor of having constructed by far, the most powerful magnets that have ever been known, and his last . . . is eight times more powerful than any magnet hitherto known in Europe."

Silliman's praise was not incidental. In Utrecht in the Netherlands, Gerard Moll had also seen Sturgeon's magnet and set about making improvements through meticulous experimentation. Increasing the number of turns of wire around the core, the Dutch professor had hoisted 75 pounds and then 154 pounds. Wasting no time, he quickly announced his discovery in the *Edinburgh Journal of Science*, and his magnet was hailed as the most powerful in the world. In large part, Henry's belated publication of his findings was in response to Moll's efforts.

Henry followed up publication of the paper with a proposal to build an even larger magnet for Yale. Silliman enthusiastically accepted the offer. The magnet Henry had in mind would have a core that weighed nearly 60 pounds and would lift an estimated 1,000 to 1,200 pounds. In the end, the device surpassed even Henry's expectations, hoisting more than a ton. The core itself was constructed of an octagonal iron horseshoe 30 inches long, a foot high, and 3 inches thick. The core was wrapped with 800 feet of copper wire divided into 26 strands.

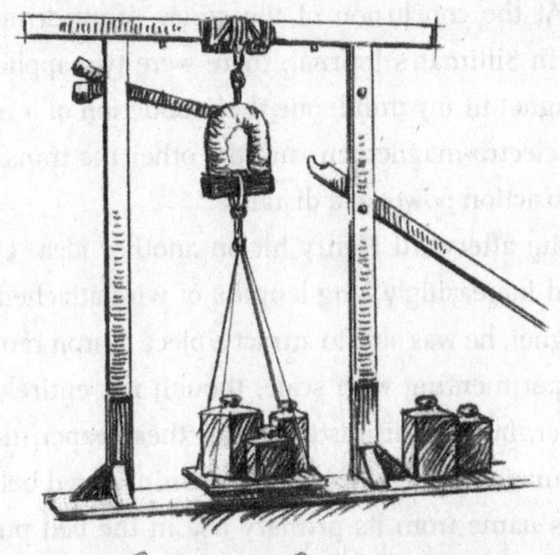

*Henry's Yale Magnet*

The battery Henry designed for the magnet was made up of concentric cylinders of copper and zinc, which he calculated offered about five square feet of active surface area. This measurement is significant not only because it reveals how batteries were still measured in terms of metallic surface area, but because Moll's less powerful magnet was also considerably less efficient, requiring some 170 square feet of surface area to power it up.

By the standards of the day, Henry's electromagnets were impressive devices. Never before had electricity been used to perform such heavy lifting, hoisting more than a man could manage without levers or pulleys—and doing it with the mysterious power of electricity generated by careful arrangement of metal and chemicals. Unlike a steam engine whose complex mechanical workings could be more or less understood through careful examination of boilers or gears, one needed to know the principles of electromagnetism and batteries to fully grasp the workings of the magnet.

It did not take a great leap of imagination to see that the power generated by the simple device would someday find use in industry. Henry himself had seen the practical potential for electromagnets, writing, "At the conclusion of the series of experiments which I described in Silliman's Journal, there were two applications of the electro-magnet in my mind: one the production of a machine to be moved by electro-magnetism, and the other the transmission of or calling into action power at a distance."

Not long afterward Henry hit on another idea. Using a small battery and increasingly long lengths of wire attached to a modest electromagnet, he was able to attract a piece of iron remotely. Again, he was experimenting with scale, though not entirely in terms of sheer power, but also in distance. For these experiments, as with his electromagnets, he used lengths of uninsulated bell wire, which received its name from its primary use in the bell pulls of household bells that summoned servants or announced visitors. A common enough item at the time, bell wire retained its name long after manufacturers had begun insulating it with a thin layer of rubber or soft plastic and its use had shifted from mechanical to electrical applications.

Soon, Henry replaced the iron chunk with a homemade device that clicked, and then he replaced the clicker with a magnetized metal bar mounted on the wall, which pivoted next to a small bell. When current ran through the wire, the magnetized rod obediently turned on its axis to ring the bell, providing an early germ of an idea for the telegraph. The device was not unlike Ørsted's compass needle that twitched when caught by a magnetic field or the primitive voltmeters that followed. In the end, Henry strung more than a thousand feet of bell wire around the walls of his classroom at the Albany Academy to test the flow of electricity.

It didn't take long before Henry's work with electromagnets caught the attention of Amos Eaton, a consultant for Penfield Iron Works in what was then known as Crown Point, New York, on Lake Champlain. The problem the firm faced was separating high-grade iron ore from lesser-quality ore. The solution Henry arrived at—similar to a cotton gin—was a wooden cylinder into which hundreds of metallic and electromagnetized rods were inserted. The high-caliber iron simply stuck to the teeth to be brushed off and sent to the furnaces for smelting. Electromagnets had found their first industrial application, and shortly thereafter Crown Point was renamed Point Henry.

THEN IN 1831, HENRY BUILT his first electric motor, powering it with a voltaic battery. "I have lately succeeded in producing motion in a little machine, by a power which I believe has never before been applied in mechanics—by magnetic attraction and repulsion," he wrote.

The "little machine" was made up of a nine-inch iron bar wrapped with copper wire and mounted horizontally so that it rocked like a seesaw. Two permanent magnets of the same polarity were mounted under each side of the bar with two thimbles of mercury attached to a voltaic cell positioned under each end of the bar. So when a wire dangling from one end of the bar dipped into one thimble of mercury, it completed the circuit and activated the electromagnet that repelled it away from the permanent magnet, forcing the other end to dip, which had the same effect. According to Henry, the machine could work at approximately 75 "vibrations a minute" for about an hour, which was as long as the battery lasted.

It was, according to historians, the first device that showed potential for electricity to perform "work." Two years later—in 1833—William Ritchie independently invented an electric motor based on the same principle in England, though his battery-powered device moved in a circular, rather than seesawing, motion. And in 1834 Thomas Edmundson of Baltimore, Maryland, modified Henry's machine to move in a circular motion. While it is likely that Ritchie had no previous knowledge of Henry's original seesaw device, after much prodding by Henry, he did eventually offer some measure of credit to the American inventor and scientist.

The little seesawing motor caused a stir. It seemed possible that electricity might someday be made to perform all manner of work, perhaps eventually even replace the steam engine. Henry remained skeptical. It was not the potential of electromagnetic power that reined in his enthusiasm; it was the limitations of batteries. For instance, the battery he built for Silliman and Yale was powerful, but that power was wholly reliant on the imperfect battery technology of the day. With the exception of the device he built for the ironworks, batteries were still largely confined to the laboratory environment and produced only limited power.

Using enough batteries, it was, of course, possible to generate sufficient electricity to power a large electric motor. Someone might take the idea, improve upon it, and increase its scale, just as Henry had done with Sturgeon's electromagnet. Such a motor could, at least theoretically, perform work, but coal was still more economical and the technology more or less perfected. Batteries were expensive and their expense, along with their size, increased with the amount of power desired.

Henry was right to remain a skeptic. Batteries were far from ready for widespread use in industry. One major obstacle was the polarization that took place within the cell, slowing the oxidation and the flow of current. For example, in a simple battery with a zinc anode and copper cathode, positive hydrogen ions released from the zinc during oxidation accumulated on the negatively charged copper to form a thin film of tiny bubbles that reduced the output. The same chemical reaction that set free the electrons also eventually blocked the flow of electricity. And, too, the electrolyte and electrodes had to be changed or cleaned frequently.

To solve the problem, scientists hit on a number of solutions, none of them particularly elegant. At the University of Pennsylvania, Dr. Robert Hare invented a battery he called a "deflagrator," basically a battery in which the plates could be easily removed from the trough to disperse the bubbles on a regular basis. Well suited for the laboratory, the deflagrator wasn't particularly practical for the outside world.

It wasn't until the 1840s that Sir William Robert Grove would solve this problem by creating a battery that included a zinc anode and platinum cathode with a porous material between them and two different acidic substances or electrolytes—sulfuric acid for the anode and nitric acid for the cathode. Grove's nitric acid battery essentially depolarized itself.

Perhaps one of the most extraordinary coincidences that saw the work of Henry and Faraday overlap was the nearly simultaneous discovery of induction in the early 1830s. Although Faraday is generally credited with the discovery, Henry took the concept a step further. Clearly not a follower of Faraday's "Work, finish, publish" philosophy, many of Henry's most important experiments exist in a volume called *Lectures on Natural Philosophy by Professor Henry* (1844), compiled by one of his students, William J. Gibson. In one particularly intriguing experiment, Henry described how a discharged or "sparked" Leyden jar in one room was picked up by a coil of wire in another room. He called the finding "induction at a distance."

According to Gibson's transcription, Henry stated,

> Hence the conclusion that every spark of electricity in motion exerts these inductive effects at distances indefinitely great (effects apparent at distances of one-half mile or more) are another ground for supposition that electricity pervades all space . . . A fact no more improbable than that light from a candle (probably another kind of wave or vibration of the same medium) should produce sensible effect on the eye at the same distance.

It was a stunning piece of scientific reasoning. However, it would take years before it was proven and further demonstrated by Heinrich Hertz and James Clerk Maxwell. Eventually it would form the basis for radio transmissions. Today, Henry receives scant credit for his findings regarding induction at a distance or transmission of electrical impulse. That is not to say he received none. Just as Volta is remembered through the volt and Ampère through the amp, levels of induction are today measured by the henry.

Like Faraday, Henry was a true believer, though he lacked the support and considerable resources of the Royal Institution. Publishing only infrequently, he eschewed patents, thinking science would progress faster without them and blaming them for holding European science back. This was the type of attitude that Franklin had adopted, but much had happened since Franklin's day.

The Industrial Revolution, which began in the mid-eighteenth century in Great Britain, was now blossoming, even in America. Artisans were giving way to larger manufacturing concerns that were adopting principles of engineering to create ever-larger and more complex production processes, and perhaps most significantly, the courts were getting better at enforcing patent laws. Much of science, once viewed as largely hobbyists' pursuit, was edging closer to the realm of commerce. The high-minded approach that had garnered Franklin praise was increasingly seen as naïve. "I did not then consider it compatible with the dignity of science to confine the benefits which might be derived from it to the exclusive use of any individual," he wrote late in life, but then added, "In this I was perhaps too fastidious."

Approached by the College of New Jersey in Princeton (renamed Princeton University in 1896), Henry wrote back saying, "Are you aware of the fact that I am not a graduate of any college and that I am principally self-educated?" In the end, it didn't matter. By 1832, with the support of Silliman at Yale, he was comfortably ensconced as a professor in Princeton and experimenting in a newly built laboratory, continuing his experiments with electricity and battery power. In 1834, he hit on the idea of creating a battery whose output could be increased and decreased at will. Using zinc encased in copper plates, each about nine inches wide and twelve inches deep, he fitted them into a slotted box arranged as one entire group of eleven and then added eight individual cells. He then connected the cells to a crank mechanism to raise and lower either the entire group of eleven or any of the individual eight cells.

From our own twenty-first-century vantage point, Henry's battery is both technically incongruous and ingeniously admirable. There is something odd about controlling electrical output with a mechanical device that resembled nothing so much as the mechanism for hoisting and lowering sails on a ship. It was also, as engineers say, an elegant solution, at least for its time.

One of Henry's more enduring discoveries was the transmission of electrical impulses over great distances through wires. In 1835 or 1836, he had replaced his wire looped around the room at the Albany Academy with wire strung between buildings on the Princeton campus. Enlisting the help of students, he significantly increased the scale of his experiments, just as he had done with Sturgeon's magnet. What he discovered is that current steadily loses its power when transmitted over long distances. To keep the current flowing at high levels from point A to point B, he engineered a device that would open and close another, secondary circuit along the way with its own smaller battery and electromagnet. The device would later become known as a "relay" and was essential to the development of the long-distance telegraph. Without relays, current moving through the wire simply became too weak to detect after a few miles.

AFTER SOME FOURTEEN YEARS, HENRY moved from Princeton to the not yet fully formed Smithsonian Institution, where he became secretary and played a pivotal role in one of the more interesting chapters in American science. Founded on roughly the same principles as the Royal Institution by James Smithson, a British subject, the Smithsonian was to be the legacy of a man who would never see the final result.

Although an early member of the Royal Institution, Fellow of the Royal Society, and enthusiastic experimenter, Smithson felt he was never accorded the full respect due a gentleman of science. The illegitimate son of Hugh Smithson, first Duke of Northumberland, and Elizabeth Macie, a lineal descendant of King Henry VII, he was accepted for his lineage but never entirely welcomed into British society because of his out-of-wedlock birth. Frustrated, he left England and set up residence in Paris on the rue Montmartre, where he welcomed American visitors.

"The best blood in England flows in my veins; on my father's side I am a Northumberland, and on my mother's side I am related to kings, but this avails me not. My name shall live in the memory of man when the title of Northumberland and the Percys are extinct and forgotten."

Apparently realizing he would never make his mark through the fruits of genius in the same way as a Newton, Cavendish, or Davy, Smithson's bid for scientific immortality was the Smithsonian Institution, which he established by leaving the bulk of his fortune ". . . to the United States of America, to found at Washington, under the name Smithsonian Institution an establishment for the increase and diffusion of knowledge among men." The fortune came to 105,960 gold sovereigns, 8 shillings, and 7 pence (and arrived in the United States wrapped in paper).

There is little doubt that Smithson had in mind something along the lines of the Royal Institution and its charter for ". . . diffusing the Knowledge, and facilitating the general Introduction of Useful Mechanical Inventions and Improvement; and for teaching, by Courses of Philosophical Lectures and Experiments, the application of Science to the common Purposes of Life."

What Smithson envisioned was significant. America still lagged far behind Europe in science and even basic industrial skills. Under British law certain skilled tradesmen deemed valuable were forbidden to emigrate. In one notable case, two brothers named Hodgson, from Manchester, camouflaged their tools as fruit trees, sending them ahead to America on a separate ship.

Yet for nine years Congress squabbled over just what ". . . for the increase and diffusion of knowledge among men" really meant, maneuvering politically and consulting with the best minds they could find, including Faraday and Henry. John Quincy Adams wanted the money to go to an observatory while others argued for a national library or a college. In the end, Congress decided on a library, museum, and art gallery. Henry proved instrumental in laying the foundation of the Smithsonian, ensuring that it would endure.

LONG BEFORE HENRY ARRIVED AT the Smithsonian, his work caught the attention of Thomas Davenport, a blacksmith from Brandon, Vermont. Born in 1802 into modest circumstances, Davenport was apprenticed at an early age. The apprenticeship apparently "took," since he went into the trade and by most accounts made a good living. However, he also maintained a lifelong habit of reading and self-improvement, supplementing his spotty education, which included just three years of formal schooling.

By most accounts there was absolutely nothing to distinguish Davenport from the thousands of other smiths who set up shop in small towns throughout New England—that is, until he happened to visit the Penfield Iron Works in 1833 and saw the industrial magnet that extracted the high-grade iron ore. It was something very much like love at first sight. The young blacksmith became entranced by the machine as well as by a simple electromagnet the ironworks had on hand. Later he would describe the magnet as " . . . an electro magnet weighing about three pounds, to which were attached two sets of cups consisting of copper, or zinc, cylinders to be set in earthen quart mugs." Unfortunately, the worker describing the machine to Davenport mistakenly called the electromagnet a battery and the battery the "cups." For years, Davenport would likewise mislabel the components until he was finally set straight.

Davenport's enthusiasm apparently got the better of him, and he offered to buy the small, four-pound demonstration electromagnet from the Iron Works. Using money intended for the purchase of iron and borrowing from one of his brothers, he eventually bought the magnet for $75.00. Probably viewing the purchase as a whim, friends and family suggested he set it up in his blacksmith's shop and charge admission to watch it in action. In time, and with a little luck, he might make the best of a foolish choice and even recoup some or all of his investment.

But Davenport had other ideas. Following in the footsteps of Franklin, he set about dismantling the device, piece by piece, to learn its secrets. With his wife taking notes, Davenport meticulously disassembled the battery and in a short time managed to duplicate the device, using (according to legend) his wife's wedding gown to insulate the tightly wrapped wires around the core. He then set about experimenting.

Within a year he had built a rotary motor that reached thirty revolutions a minute—though some accounts place the number much lower. It was not by any stretch of the imagination a very practical piece of equipment. It was not even wholly original. Henry and Faraday had both created small motors. Davenport, the most impractical of practical men, saw in the little device the potential for industrial use.

The goal of creating a practical piece of equipment must have seemed tantalizingly close for the Vermont blacksmith. After all, he had witnessed Henry's magnetic device at the ironworks, knew that electromagnets were capable of performing labor—lifting more weight than even a grown man could manage—and had a proof of concept in the little motor. What more could a young, ambitious inventor desire? It was only a matter of applying labor and brain power to the problem and somehow bringing these disparate elements together in a single device.

All but abandoning his blacksmith trade, he stubbornly kept at the task, continuing to refine his little motor to the exclusion of almost everything else. Mocked by his more sensible neighbors and with his business falling into neglect, he asked his local pastor's advice and was told, "If this wonderful power was good for anything, it would have been in use long ago." Ignoring the counsel, Davenport forged on.

Like Faraday and Henry, Davenport was a true believer, though it genuinely seems that his enthusiasm far exceeded his grasp of the scientific principles. In the historical record, he embodied one of the most enduring of all American myths: the lone inventor of humble beginnings tinkering away in the homey isolation of his workshop. Unfortunately, science does not particularly favor mythologies, even compelling ones that nurture our sense of infinite possibilities.

Davenport did eventually venture out of his workshop in search of funds to continue his research and meet with those who could offer technical advice. At Middlebury College, he visited Professor Edward Turner. He began reading scientific journals, including Silliman's, where he discovered Henry's work. Unlike Henry, he did take out a patent on his device and in due course ended up forming a company and issuing stock in an attempt to fund his research as well as extend patents to Europe. He made stops at the Rensselaer Institute in Troy, New York, in Saratoga Springs, New York, and in Boston in search of financing and then at the Franklin Institute in Philadelphia, seeking advice and financing. Finally, he arrived at Henry's doorstep in Princeton.

The legendary scientist did offer some kind words and advised the young inventor to continue his search, though he suggested building the device on a smaller scale. A smaller version, he reasoned, would not only prove less expensive to produce; defects in a smaller, experimental piece of equipment would be looked upon more charitably by potential investors than those in what appeared to be a completed industrial version.

However, in a letter to Silliman following his meeting with Davenport, Henry seemed decidedly less enthusiastic toward the blacksmith, writing, "I felt considerably interested in the welfare of the Inventor and with friendly motives advised him to abandon the invention."

Of course, Davenport did no such thing. Obsessed, he kept

plugging along, scraping up funds where he could and garnering press coverage in hopes of luring in more investors. In the *New York Herald* in April 1837, the headline read: "A Revolution in Philosophy: Dawn of a New Civilization." Unlike Henry, journalists of the day had only to see the small motor in action to be convinced of its eventual use. After a detailed description of Davenport's motor, the unnamed journalist wrote, "There can be no doubt, in our mind, but the days of steam power, and animal power, and water power, are gone forever. This is no idle vision—no fancy's sketch."

Davenport managed somehow to generate enough power from crude batteries to turn a lathe and later apply the motor to a miniature electromagnetic railroad car that he used for demonstrations, the first electromagnetic player piano, and even a printing press on which he published an ill-fated newsletter called, naturally, *The Electro-Magnet.* But unable to raise sufficient funds, the projects all but vanished, and Davenport—as heroic as he was—became a small footnote in history.

PROGRESS WAS ALSO MOVING APACE in Europe. Nicholas Callan, an Irish priest who was a contemporary of Henry's, became fascinated with electricity while studying in Rome. Appointed to the chair of natural philosophy at Maynooth College in the 1820s, he began a series of experiments that closely paralleled Henry's. At one point he built what was said to be the world's largest battery by joining more than 500 cells and 30 gallons of acid. The battery, according to reports at the time, could hoist over two tons.

And, in one of those strange instances in the history of science, his most stunning invention doesn't bear his name. In 1836, Callan discovered that by running a current through a couple of turns of heavy wire around an iron core situated near a smaller core with many turns of fine wire, the current was increased in the second coil. What he had discovered was the induction coil—a device capable of efficiently increasing current. However, unappreciated at

Maynooth—where religion rather than science was central to the curriculum—Callan's invention was soon forgotten, and Heinrich Ruhmkorff, a German-born instrument maker in France, took the credit and put his name on the device.

DESPITE DAVENPORT'S APPARENT FAILURE, MORE electric motors began to emerge. In 1838, the Scottish inventor Robert Davidson built a wide variety of electrically powered devices, including an electric-powered locomotive, which he called a "Galvani." It chugged along at a steady four miles per hour but was calculated to be forty times more expensive to run on its zinc and copper batteries than on coal.

Then, in 1839, Moritz Hermann Jacobi, a German scientist working in Russia, built a battery-powered boat at the behest of Tsar Nicholas I. Twenty-eight feet long and seven feet wide, it featured multiple paddle wheels powered by more than 300 Daniell cells (some accounts list the batteries as Grove cells) and a one-horsepower electromagnetic engine. But in the end, the batteries were simply too heavy and kept the boat from attaining anything near reasonable speed. The experiment was a technical success, but a practical failure, though later in the nineteenth century batteries would power launches up and down the Thames. In 1851, the American Charles Page received a grant from Congress to build a train powered by a hundred Grove batteries. Though marginally faster than Robertson's train, it still proved unreliable because of problems with the circuitry.

Though still expensive and exotic, battery technology continued to improve almost exclusively through trial and error. And with the ability to produce a steady and relatively long-lasting electric current from newly designed batteries—such as the Grove nitric acid battery—it was only a matter of time before some widespread technology came of it.

In Germany, Robert Wilhelm Eberhard Bunsen was also hard

at work on batteries in the 1850s. A chemist by training, he is best known today for the Bunsen burner, which he designed for the laboratory at the University of Heidelberg to replace the ad hoc collections of oil lamps that had been standard in labs. Though largely forgotten today, his Bunsen cell or Bunsen battery marked a significant step forward in battery design. By substituting a carbon rod for the much more expensive platinum cathode in a Grove battery, he lowered the cost of batteries forever. For the first time, even the humblest of amateur inventors and less-well-endowed laboratories could afford a reliable power source. And batteries—a source of the mysterious electrical power—were now within the reach of industry.

Those who predicted that electricity would first find its place in physical labor with electric motors proved equally as wrong as those who believed electricity useless or simply an interesting novelty. Given the state of the science and technology, electrical power worked best to power small devices. So, it was in communication—the telegraph—that electrical power found its first widespread application.

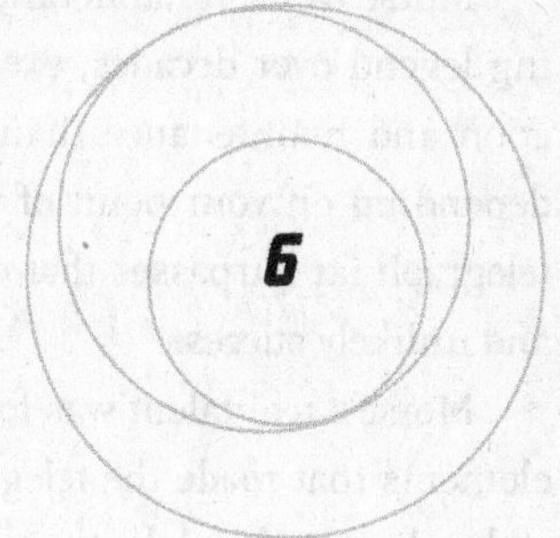

# What Hath God Wrought?

*"You can't throw too much style into a miracle. It costs trouble, and work, and sometimes money; but it pays in the end."*

*—Mark Twain,*

A Connecticut Yankee in King Arthur's Court

As the history book legends have it, Samuel Finley Breese Morse defied all doubters and stretched the boundaries of technology with his invention of the telegraph and the code that went with it. It was Morse, or so we are taught, who led the charge in the conquest of distance and united America from coast to coast with the humming, pulsing wires of his invention. What the jurist Oliver Wendell Holmes described as " . . . a network of iron nerves which flash sensation and volition backward and forward to and from towns and provinces as if they were organs and limbs of a single living body."

Morse himself handcrafted and diligently nurtured this appealing legend over decades, exerting even more effort toward its creation and maintenance than he put into the telegraph. However, depending on your point of view, the true story of Morse and the telegraph far surpasses that of his own fictional account for drama and unlikely success.

Morse's real talent was for drawing together all of the disparate elements that made the telegraph possible. And, too, he was in the right place at the right time. The scientific heavy lifting of experiment after experiment that formed the basis for the telegraph was mostly completed by the time Morse arrived on the scene.

What was missing was someone to refine the basic mechanics and coordinate the organizational components. In the corporate parlance of the twenty-first century, Morse played the role of project manager, attracting and then coordinating the expertise, funding, publicity, and even political lobbying to bring a functioning telegraph to life. This was no small thing in an age that had not yet come to fully trust technology whose mechanism could not be seen in the gearworks and boilers that powered the Industrial Revolution.

IF MORSE'S TECHNOLOGY WAS NEW, the concept itself—rapid communication over long distances—was centuries old. Instantaneous communication had long been a potent concept, even when it existed solely as a myth. As far back as the sixteenth century there were rumors and whisperings across Europe of a magical device that used "sympathetic needles," similar to a compass, to communicate over great distances. A complete description appeared in the book *Prolusiones Academicae* penned by the Italian Jesuit academic Famiano Strada in the seventeenth century.

The first actual telegraph didn't use electricity at all. Designed by Claude Chappe and his brother, René, in France, the system was made up of a series of signal towers with large semaphore-like mechanical arms that could be raised and lowered to spell out words. Although it's not often noted, Chappe had experimented with an electrical form of communication using Leyden jars, though little came of it. The technology just wasn't up to the job, so he switched his attention to visual communication. Originally Chappe thought to call the system a "tachygraphe" from the Greek words for "fast" and "writer." In the end, a friend, who happened to be a classics scholar, suggested the more descriptive "telegraphe" for "far writer."

The towers, which were arranged every few miles, eventually crisscrossed France, their controls of levers and pulleys artfully designed by the clockmaker Abraham-Louis Breguet, whose name continues on luxury watches and who would also later enter the telegraph marketplace as a manufacturer of electromechanical receivers. Though cumbersome and labor intensive, the signal towers were a communications breakthrough. The 1797 edition of the *Encyclopædia Britannica* paid homage to the Chappe brothers' invention, noting, "The capitals of distant nations might be united by chains of posts, and the settling of those disputes which at present take up months or years might then be accomplished in as many hours." The optimism expressed in the encyclopedia's entry was as genuine as it is enduring. Two centuries later similar sentiments would be voiced amid the growing popularity of the Internet.

WITHIN SEVERAL YEARS, OPTICAL TELEGRAPH systems, as they were eventually known, began to spring up across Europe. Napoleon was an enthusiastic supporter, ordering the construction of the system to extend to Boulogne, and then launching a study to assess the feasibility of signaling across the English Channel in preparation for an invasion of England. By the mid-1830s, all of Europe was dotted with towers of varying designs that worked reasonably well—a big step forward.

Today we measure communication in near real time—that is the time it takes to speak into a telephone or write and send an e-mail or text message. But throughout most of human history communication time was measured in terms of transportation, travel time. That is to say, the time it took to communicate was defined by the speed of horses or ships and factored in a large number of variables, such as weather and even the reliability of the messenger. Delays in communication or miscommunications were so common they became a standard plot device in drama. Shakespeare's characters frequently used untrustworthy or slow-moving messengers that conveniently shaped the plot. *Romeo and Juliet* would have been a much different play if the star-crossed teens had had access to text messaging or cell phones.

As transportation improved along with infrastructure—faster carriages, faster ships, better roads, more accurate navigation—so did the time it took to communicate. However, the idea of communicating over hundreds of miles within a few hours via the Chappe brothers' invention must have seemed revolutionary, an engineering miracle, and a point of national pride not unlike the Roman viaducts or America's railroads.

The technological seeds that would render the Chappe brothers' towers obsolete were planted as early as 1816 when Sir Francis Ronolds (sometimes spelled Ronalds), an English meteorologist, used a friction generator to send an electrical impulse down a wire to move a pair of suspended pitch balls—a crude voltmeter or electrometer. He offered his idea to the Royal Admiralty, which summarily rejected it. Electricity was still the stuff of natural philosophy as far as the military was concerned. And, of course, there was Henry, who had been amusing his students by sending electrical current through wires to ring a bell since his early days at the Albany Academy. Among Henry's inventions, the relay—a secondary electromagnet and battery to pass current along the line—would prove essential to the development of the telegraph.

Then there was a mysterious letter that appeared in the February 17, 1753, issue of *The Scots* magazine signed only "C.M." The letter detailed an eerily prescient plan for an electrical telegraph system that included poles and overhead wires, one for each letter of the alphabet. And then, in the 1790s Agustin de Betancourt, a businessman and engineer, ran a line made up of more than seventy wires between Madrid and Aranjuez, while Carl Friedrich Gauss, the German mathematician, set up a simple telegraph system across the rooftops of the University of Göttingen as early as the 1830s with a device that moved a needle to the left or right. There were other experiments serving to varying degrees as proof of concept.

It was into this environment of proven science and early prototypes that Morse stumbled. As the legend goes, he was a struggling artist struck by inspiration during a return voyage from Europe. The first part, of course, is true. Morse was an artist, and somewhat struggling. There could also be no denying that he was passionate about his art. In an early letter to his mother, he immodestly wrote, "My ambition is to be among those who shall revive the splendor of the 15th century, to rival the genius of Raphael, a Michael Angelo [*sic*], or a Titian; my ambition is to be enlisted in the constellation of genius which is now rising in this country."

To say the least, Morse was a man of oversized ambition as well as fierce, often fanatical patriotism. That he saw himself in the center of a second Renaissance in early nineteenth-century America was not out of character. His masterpiece, called the *House of Representatives*, was an immense canvas weighing more than 600 pounds. The huge painting, according to Morse, was intended to show " . . . a faithful representation of the national hall with its furniture and business during the session of Congress." He toured the painting, which could be described as a very early version of C-Span, charging admission to see it, though in the end it turned out to be a modest failure when the expected crowds did not materialize.

BORN AS HE WAS IN 1791, in the shadow of the American Revolution and a little over a mile from Franklin's birthplace, it is not surprising that Morse was very much pro-American in his views.

However, the influence of his father, Jedediah, is more than likely responsible for pushing that sensibility to extremes. An evangelical Calvinist preacher, Jedediah harbored dark suspicions of secret conspiracies aimed at destroying America. He preached sermon after sermon on the dangers of Catholicism, Masons, the Illuminists, and the dreaded French imperialism.

If Morse inherited his passions as well as his prejudices from his father, what he lacked was an outlet. Unable to break into the lucrative portrait business or create an American Renaissance, Morse was ripe for a new project into which to pour his considerable energies when the new technology of electromagnetic telegraphy captured his imagination. He had already dabbled in inventing. In 1817, along with his brother, he came up with a flexible water pump that received some notoriety along with a patent, if little financial success.

Bouncing from New York to Europe to Mexico and back to Europe, Morse encountered Dr. Charles Jackson on one of these voyages sailing back home from England aboard the *Sully* in 1832. A twenty-eight-year-old Harvard-educated Boston physician and amateur scientist, Jackson was returning from Europe where he had been studying geology. Among his rock samples, the good doctor also had on board a small electromagnet and battery, which he demonstrated. Morse, who was forty-one at the time, was intrigued enough to begin experimenting with electricity.

However, Morse was easily distracted. When he reached New York, he secured himself a professorship at New York University, kept up with his painting, helped establish the National Academy of Design, and began one of the strangest phases of his life, that of a social activist and politician.

As Morse and a few others saw it, America was in grave danger from immigration. By today's standards, and even those of his own day, Morse came out on the politically incorrect side of nearly every political and social debate. A fierce nationalist, he wrote prolifically on the threat from immigration, even going so far as to edit a manuscript called *Confessions of a French Catholic Priest.* Published in 1837, at a time of rampant "antipapist" sentiment, it's an extraordinary lurid piece of hateful propaganda. Filling some 250 pages, the book is packed with tales of murder, sexual depravity, and bizarre practices, such as drinking mysterious elixirs fermented from water lilies. Later, he would pen a pamphlet condemning the abolitionists, describing them as "demons in human shape."

Thoroughly engulfed in the Nativist movement, Morse ran for New York City mayor in 1836 as the candidate for the Native American Democratic Association, whose core platform was a particularly virulent form of anti-Catholicism. He was thoroughly beaten, getting just 1,500 votes to his Democratic opponent C. W. Lawrence's 16,101.

Then, in 1837, word reached him—through his brother in the newspaper business—that a pair of Frenchmen named Gonon and Serval were demonstrating a device that could send messages instantaneously across great distances. Morse, who had been toying with an electric telegraph off and on since returning from England, didn't realize the two inventors were just updating the optical Chappe system, though it did prompt him to pay more attention to the technology and learn that an electromagnetic telegraphic concept was being used throughout Europe in isolated instances with varying degrees of success.

ONE OF THE MOST PROMISING and successful of these early efforts was that of Charles Wheatstone and William Fothergill Cooke. Wheatstone, the son of instrument makers and an inveterate tinkerer, had gained some local fame with an instrument he called the "enchanted lyre" or "aconcryptophone," which seemed to sound like a variety of instruments. If nothing else, it was a neat trick. The cord from which it hung was actually a hollow rod that transmitted the sounds speaker-tube style of other instruments played in a different room. The lyre itself was a type of acoustic amplifier. Later Wheatstone would follow up this invention with the much-maligned accordion.

Cooke, who left military service before retirement age, was scraping out a living of sorts by making anatomical wax models for use in medical schools. Like Morse, he was a man very much in search of the main chance and the fortune it would bring.

Wheatstone and Cooke hit on the idea of the telegraph almost simultaneously in the 1830s and eventually decided to join forces. Introduced by Peter Roget of thesaurus fame, who conducted a series of gentlemanly electrical experiments himself, the pair made an odd team. Fothergill had been inspired by a lecture on Baron Pavel Lvovitch Schilling—a Russian diplomat who invented a crude electromagnetic telegraphic system in the 1820s. The unit essentially used a voltmeter to signal from place to place, much like Henry's early experiments. After much effort Schilling finally convinced Tsar Nicholas I to construct a telegraph network, then promptly died before construction began, and the idea was scrapped.

However, one of Schilling's voltmeters fell into the hands of a professor in Heidelberg as a curiosity, and it was this unit that Cooke witnessed in use during the lecture. Cooke immediately set about designing his own version, which used three needles and six wires along with an unwieldy code.

Wheatstone, who was by turns intolerably arrogant and painfully shy, had designed a simple telegraph with six wires that activated five separate dials on a beautiful diamond-shaped face. He also had a stockpile of some four miles of wire. Only by teaming up were they able to get their design off the ground, though the pair feuded for years. Cooke, the public face of the duo, treated Wheatstone as a junior partner. Wheatstone, for his part, was insistent that he receive credit for the device, arguing over whose name appeared first on paperwork.

Whatever their personal problems, by the 1830s the pair were granted a patent for their system—the first for electrical transmission. One of the first practical uses for their telegraph was through a mile of wire that stretched between the Euston and Camden Town terminals in London to signal arrivals and departures. The experiment was a success—even the public embraced the new technology, primarily because it made obsolete the piercing whistles and drums that had been used previously.

Cooke and Wheatstone's telegraph would eventually gain prominence in the United Kingdom, chiefly because of the comprehensive patent they filed. In fact, their patent in America, dated June 10, 1840, beat Morse, whose filing is dated June 20, by ten days.

SPURRED INTO ACTION, MORSE LAUNCHED a publicity campaign, first by getting his journalist brother to announce his invention, and then by polling other passengers aboard the *Sully* to establish a time line for his invention. He became obsessed with beating the Europeans. Only one thing stood between Morse and what he saw as his rightful place in the history books: his telegraph system was overly complex and less efficient than systems Henry had employed years before Morse even boarded the ship. That is not to say the system didn't work; it did, but only up to a point. That point was about forty feet before the signal dropped precipitously.

To help solve his problems, he visited Henry, then at the College of New Jersey in Princeton, who freely shared his thoughts on the subject, which included increasing the power of the batteries as well as the concept of relays. He also enlisted the assistance of Leonard Gale, a professor of chemistry at New York University. The problem, as Gale saw it, was simple: Morse was using the wrong kind of battery and magnet. After substituting the single cup battery of zinc and copper for larger, more efficient batteries made up of forty cups or cells, Gale set to work on Morse's electromagnet.

Morse, who had not kept pace with the latest scientific advances, had built an electromagnet closer to Sturgeon's loosely wrapped model than to Henry's tightly wrapped design. By simply increasing the number of turns of wire around the core, Gale was able to substantially increase the magnet's power. These two design enhancements—both widely known in the scientific community for years—boosted the signal's range from 40 to about 1,700 feet.

Although much improved, the system was still not practical in any commercial sense. Morse could, like Henry, string wires around his artist's studio or between the buildings of New York University, but he had much larger plans. By 1837 he arranged a public demonstration followed up with a letter-writing campaign asking for government funds. His timing could not have been worse. The speculative bubble on Wall Street burst in May of that year, causing the Panic of 1837, closing nearly half the banks in the country and drying up funds. To sway public opinion, Morse stepped up his publicity campaign. Articles about his device were quickly published in magazines intended not for scientists, but for the general public, such as *The Journal of Commerce.*

He also enlisted the help of one of his students, Alfred Vail, whose family conveniently owned the Speedwell Iron Works in New Jersey. Along with Vail came the young man's mechanical expertise, the resources of the tool-and-die company, and the much-needed funds the well-to-do Vail family could provide. The first battery the partnership produced was a lovingly designed Grove-type unit housed in a cherrywood box lined with beeswax. The little chemical power plant was downright elegant and so was the wiring, which they insulated with material used by milliners for hats.

Vail also possessed a better sense of industrial design than Morse, and within a very short time he had streamlined the telegraph key or register. The young man was also clever with code. Rather than adopt the massive dictionary comprised of entire of words Morse had labored to assemble, he settled on a simple binary code of dots and dashes representing letters of the alphabet. In researching his code-making Vail consulted with the typesetters at a local paper to see which letters appeared most often. It was these frequently occurring letters for which he saved the simplest groups of dots and dashes. To represent the letter "e," for example, Vail assigned a single dot, while the less frequently used "q" was given the more complex combination of two dashes, a dot, and another dash.

This code, of course, eventually became known as Morse code rather than Vail code. It is not to Morse's credit that he would invariably refer to Vail as his "mechanical assistant."

MORSE GOT HIS BIG BREAK when a proposal went before Congress to construct a series of Chappe-style signal towers between New York and New Orleans. Taking his simplified electromagnetic telegraph system to Washington, he demonstrated the device by setting up the sending and receiving stations a few feet apart. The lawmakers, who couldn't quite grasp the technology, were profoundly unimpressed. Undeterred, Morse decided to try his device in Europe, where Wheatstone's system was catching on, then returned to the United States and arranged for another demonstration, this time stringing wires between two committee rooms of the Capitol building.

The revised demonstration got the lawmakers' attention, but Morse still needed to lobby hard, soliciting letters from well-known scientists. In one of these letters, Henry called for the construction of the telegraph as a matter of national pride in much the same way Davy had solicited funds for his oversized battery and similar to the way President John F. Kennedy would rally a nation around the space program. Henry's letter argued for the project's funding not only as a practical matter of communication, but also to "advance the scientific reputation of the country" and to ". . . be furnished with the means of competing with his European rivals."

Morse was tireless in his promotion, launching demonstrations of his electromagnetic telegraph for anyone who might help him get the project off the ground. In one of the stranger instances, he joined efforts with Samuel Colt, of Colt revolver fame. The inventor of the "gun that won the West," just twenty-eight at the time, was working as a research scientist for the Department of the Navy, trying to employ electricity as a weapon.

What Colt had in mind was a waterproof battery, called a submarine battery, capable of exploding mines as a means of harbor defense. Though working on the battery in the strictest secrecy, he eventually gave four public demonstrations in Washington and New York City. Much later, following on the heels of Morse, he would switch his focus to telegraphy, constructing his own short-lived telegraph system from Coney Island to Manhattan to announce ship arrivals.

In 1842, Morse attempted to send a signal across New York Harbor, but the experiment failed because of a weak battery. The same day, Colt launched his own far more dramatic demonstration. With a crowd of some 40,000 gathered along the shoreline of lower Manhattan, he used an electrical charge to blow up a ship, curiously named the *Volta*, in New York Harbor. There was nothing particularly unique about the demonstration except, perhaps, scale. Davy had ignited a small amount of gunpowder on stage during his demonstrations using a Leyden jar, while Volta had performed the same trick for Napoleon with a battery. Still, Colt's pyrotechnics must have been impressive.

Decades later, in Mark Twain's 1889 classic, *A Connecticut Yankee in King Arthur's Court*, the protagonist, Hank Morgan, perhaps taking a page out of Colt's explosive promotion, performs a similar feat by blowing up a building via an electrical charge.

> We knocked the head out of an empty hogshead [cask] and hoisted this hogshead to the flat roof of the chapel, where we clamped it down fast, poured in gunpowder till it lay loosely an inch deep on the bottom, then we stood up rockets in the hogshead as thick as they could loosely stand, all the different breeds of rockets there are; and they made a portly and imposing sheaf, I can tell you. We grounded the wire of a pocket electrical battery in that powder, we placed a whole magazine of Greek fire on each corner of the roof—blue on one corner, green on another, red on another, and purple on the last—and grounded a wire in each . . .

. . .

TWAIN'S MEMORABLE WORDS, "YOU CAN'T throw too much style into a miracle," accurately summed up the view of technology in Morse's time. It was a slogan that Morse, as well as Colt, could have adopted as their own. In Twain's book, the public, along with royalty, is quick to substitute one belief system for another. For Morse, the task was not nearly as daunting. He simply needed to convince the Washington lawmakers to write a relatively large check.

Still, this was not as easy as it seems. The technology was baffling and downright suspect to the lawmakers. During the last debate on the House floor in early March 1843, Representative Cave Johnson of Tennessee famously ridiculed the proposal, joking that they start funding mesmerism along with the "Electro-Magnetic Telegraph." Still, the measure passed, and Morse received his money, but just barely and without much enthusiasm from Congress. Seventy congressmen did not vote at all "to avoid responsibility of spending the public money for a machine they could not understand." In the end, Morse received $30,000 for an experimental line to run between Baltimore and Washington, about forty miles.

The idea was to lay the wires along an existing railway line underground, a system abandoned halfway through the project in favor of overhead poles. This made sense not only because it required permission from just a single entity, rather than dozens of homeowners and businessmen, but from an engineering standpoint as well. The railroad connected the two cities with a direct path that had already been surveyed and cleared. Later, as telegraph lines began to stretch across the country, the railroads would prove particularly valuable to the effort. Many of the smaller rural communities, though linked by rail lines,

MORSE TELEGRAPH KEY

were not linked by direct roads. Railroads would also provide office space at local depots as well as personnel for telegraph operations.

Eventually the railroad, the Baltimore & Ohio Railroad Company, agreed to participate on the condition that it had access to the system, if the contraption actually worked, which was far from a certainty. Then, covering all its bases, the railroad's lawyers added a clause to the contract that clearly stated the telegraph would function "... without embarrassment to the operations of the company."

In America, telegraphs would remain closely linked with railroads just as in England they maintained close ties with the postal service. This would prove a mutually beneficial relationship for decades as the railroads continued to expand their operations and needed a means to communicate while telegraph companies wanted the surveyed, single-owner land between cities.

Even after the funding was approved, doubts persisted in Congress. Fearing the inevitable scandal that would accompany a failure of Morse's technology or his exposure as a charlatan, the lawmakers appointed John W. Kirk as an observer. If Morse turned out to be a fraud or the technology flawed, Kirk would know soon enough, and Congress could get out in front of any accusations of chicanery.

The first experimental line was an awkward thing in terms of power, employing huge Grove batteries with eighty cells at the sending and receiving ends, though Morse later managed to cut the number down considerably. The relays that carried the transmission along the forty miles were also oversized. According to one account, they weighed some 150 pounds and were housed in three-foot-long wooden boxes two feet wide. In concept, these relays were relatively simple things. With each electrical burst from the telegrapher's key, the relay's own electromagnet opened a new circuit powered by a battery inside the box. That charge would send a fresh burst, duplicating the telegrapher's original, along to the next relay.

The relay concept, originally thought up by Henry, was an entirely different way of dealing with electrical transmission over long distances. With relays it wasn't necessary to transmit a powerful signal along a thousand miles of uninterrupted wire. All that was needed was to send a signal to an electromechanical relay—a matter of a few miles. By adding relays, the range of the telegraph became limitless. "If it will go ten miles without stopping," Morse would later say, "I can make it go around the globe."

IT IS TRUE THAT MORSE famously transmitted "What hath God wrought!" a biblical quotation from Numbers 23:23, as the first telegraph message on the line from the Supreme Court chambers to Mount Clare Depot in Baltimore on May 23, 1843. That is to say, it is technically true. However, the first actual transmission over the line was significantly less poetic or memorable. As a practical demonstration for Kirk, Vail transmitted the names of the Whig National Convention's nominees some three weeks earlier. The line had not yet been completed, falling around fifteen miles short of Baltimore, but the demonstration was enough to convince Kirk of the telegraph's usefulness after the train carrying the same names arrived more than an hour later.

News of Morse's success with his "lightning line" spread quickly, but the telegraph did little immediate business. When the first public telegraph line from Baltimore to Washington was opened, the total receipts came to about a dollar the first week. Offered for sale to the government for $100,000, lawmakers turned it down as a nonmoneymaker, and later, when a telegraph office opened in New York City, Morse charged admission to watch the telegrapher at work as a way to subsidize the company.

Still, the battery—electricity itself—had reached a turning point. With the telegraph's usefulness, if not profitability, proven beyond all doubt, batteries entered the world of commerce and industry.

Very quickly, telegraphy grew into a major concern. Along with the railroads, telegraph companies would be among America's first large corporations. As it turned out, the government and nearly everyone else misjudged the potential of the technology. Within two years of the first transmissions, a large roster of independent companies had strung up about 2,000 miles of cable, and by 1850 there were said to be some 12,000 miles of wire crisscrossing the countryside and cities. The electromagnetic telegraph caught on in Europe with hundreds of miles of wire stretching out in spokes from the major cities. France, notably, seems to have gotten a late start simply because it had such an early start with its Chappe optical telegraphs. The system, more sophisticated and widespread than any in Europe, worked just fine as far as many of the French were concerned. As an interesting historical note, France did eventually get into telegraphy with a system at least partially designed by Louis-François Breguet, grandson of Abraham-Louis Breguet, who had designed the mechanism for the optical telegraphs. Called the "French telegraph" or "Breguet telegraph," the system remained in use for years in both France and Japan.

It didn't take long before businesses seeking an advantage over their competitors began to rely on the telegraph. Just as personal and portable devices have changed expectations for business communication in the twenty-first century, the telegraph very quickly set the standard in the mid-nineteenth century.

Codebooks flourished as businesses sought to keep their private affairs secretive. Within a year after Morse's famous message traveled forty miles, former Congressman Francis O. J. Smith, Morse's lawyer and sometime promotional agent (who just happened to have headed the committee that approved the experimental line), published a commercial codebook called *The Secret Corresponding Vocabulary: Adapted for Use to Morse's Electro-Magnetic Telegraph.*

Unlike railroads or shipping lines, which required vast amounts of money to launch, virtually anyone could enter the telegraph business. With companies multiplying at an exponential rate and wires unspooling across the continent at breakneck speed, the situation grew chaotic. As more and more companies sprang up, the quality of service quickly dropped, even as competitors began slicing away at the profit margins. To compete, companies began stringing ever more wire, rapidly expanding the networks in an attempt to gain market share.

Finally, companies began consolidating, first regionally, and then nationally. In the 1850s, businessman Hiram Sibley saw the mood right for mergers and formed Western Union out of the New York and Mississippi Valley Printing Telegraph Company, founded a few years prior. His plan was simple: buy up and merge all the struggling telegraph companies he could find. "This Western Union seems to me very like collecting all the paupers in the State and arranging them into a union so as to make rich men of them," quipped a man of finance.

Of course, Western Union eventually succeeded for a variety of reasons, the most interesting of which is a nineteenth-century version of Metcalfe's Law. Named for Robert Metcalfe, an engineer at Xerox PARC and coinventor of Ethernet, the law simply states that a network increases in value as more users (or communicating devices) are added to it. Stated another way, one electromagnetic telegraph was useless, two marginally better, and six better still. Specifically, a network's value is proportional to the square of the number of devices or users.

Western Union would eventually link thousands of telegraph keys as a web of disparate lines were joined into a seamless and seemingly endless network. Economies of scale kicked in too with the standardization of equipment, including batteries and the chemicals they required, boosting the profit margin of each message sent.

The most common batteries employed were the zinc-platinum Grove cells that used nitric acid. They differed significantly from our batteries today. Very much a piece of industrial equipment, their upkeep required special training. Thick in-house technical manuals as well as books intended for the general public included pages and pages on telegraph batteries, their different components, specifications, and proper maintenance.

. . .

ALONG THE WAY SOMETHING VERY interesting began to happen. A subculture arose among telegraph operators, much as it had around swifts (typesetters) in the Victorian era and tech support personnel today. Abbreviations arose for common messages, denoting dinner breaks and chess moves. According to some linguists, the popular expression "okay" is derived or spread by telegraph slang for "Open Key Prepare to Transmit." The term "hams" for new operators would later transfer to amateur radio operators. During slow periods, telegraphers shared jokes and gossiped long distance. There were long-distance feuds between the part-time telegraph operators in rural communities, who also sold tickets and checked freight at local train depots, and the professional big-city telegraphers.

Later they would call themselves a "brotherhood." And powering it all were the batteries, hundreds of thousands of them placed in every telegraph office and relay station around the country.

The Scottish mathematician, physicist, and hobbyist poet James Clerk Maxwell, wryly captured the mood of the telegraphers and the new technology in his poem "Valentine by a Telegraph Clerk."

*The tendrils of my soul are twined*
*With thine, though many a mile apart.*
*And thine in close coiled circuits wind*
*Around the needle of my heart.*

*Constant as Daniel, strong as Grove.*
*Ebullient throughout its depths like Smee,*
*My heart puts forth its tide of love,*
*And all its circuits close in thee.*

*O tell me, when along the line*
*From my full heart the message flows,*

*What currents are induced in thine?*
*One click from thee will end my woes.*

*Through many a volt the weber flew,*
*And clicked this answer back to me;*
*I am thy farad staunch and true,*
*Charged to a volt with love for thee.*

On July 4, 1861, work began on America's transcontinental line, which carried the wires westward from Omaha, Fort Laramie, and Salt Lake City to San Francisco, linking both coasts. Subsidized by the government with $40,000, it was estimated that the project would take two years from start to finish. It was completed in four months, a full eight years before rail lines connected both coasts. The new network not only linked a sprawling continent, but definitively uncoupled communication from travel time. This was more than simply another instance of the death of distance. It was a way to efficiently govern a nation whose cities were sprawled across a continent of once seemingly insurmountable size.

The pony express, which still occupies a cherished place in America's mythology, was actually a financial disaster, in large part because of the telegraph. Launched in April 1860, the relay mail service that spanned a continent couldn't compete with the telegraph, which opened for business in October 1861. Within weeks, the pony express, which carried a letter coast to coast in ten days, was closed down after losing money for its backers.

Amid the rapid expansion, the general public could not read enough about the miracle of electromagnetic telegraphy, though not everyone was so enamored of the new technology. Henry David Thoreau sourly dubbed it "an improved means to an unimproved end." And in Hawthorne's classic *The House of Seven Gables* (1851), two characters debate the relative merits of the new technology against a backdrop of superstition and witchcraft.

> "Then there is electricity,—the demon, the angel, the mighty physical power, the all-pervading intelligence!" exclaimed Clifford. "Is that a humbug, too? Is it a fact—or have I dreamt it—that, by means of electricity, the world of matter has become a great nerve, vibrating thousands of miles in a breathless point of time? Rather, the round globe is a vast head, a brain, instinct with intelligence! Or, shall we say, it is itself a thought, nothing but thought, and no longer the substance which we deemed it!"
>
> "If you mean the telegraph," said the old gentleman, glancing his eye toward its wire, alongside the rail-track, "it is an excellent thing,—that is, of course, if the speculators in cotton and politics don't get possession of it. A great thing, indeed, sir, particularly as regards the detection of bank-robbers and murderers."

The debate, which seems quaint by today's standards, can with very little editing take on new significance, becoming a discussion on the Internet, the cell phone, or any one of the new technologies that have entered our lives over the past few decades. Then, as now, the entire world was rapidly changing. Financial news could be transmitted quickly, companies began to expand, opening branch offices, while train traffic moved with more precision.

Science also benefited. During the government-funded United States Coast Survey to measure longitude, astronomers relayed their time signals between observatories by telegraph for more accurate readings.

There were also problems. America ran on local time. The death of distance caused havoc in commerce when it came to keeping accurate schedules and conducting business long distance. The world had become split between the near instant communication of ideas and the physical world, which was (and is) very much subject to travel time. For instance, railroads were expanding nearly as rapidly as telegraph lines after 1840. By some accounts, rail lines increased more than tenfold in just a few years. Pushing west, they struggled to reconcile their schedules to the vast array of local times then commonly in use across America. An hour's train ride from east to west could throw off a traveler's pocket watch. And worse, back at the railroad company's headquarters the efficient allocation of resources and scheduling made possible by the telegraph were in danger because of the jumble of local time zones.

Adding to the confusion in America, rail lines took an ad hoc approach to the problem, often using the corporate headquarters' time as accurate. In the 1840s guides to train and steamship schedules were published to correct the problem, but they often only added to the confusion. Although usually technically accurate, differences of fifteen minutes between relatively nearby cities quickly became unwieldy for businesses and proved disastrous in 1853, when two trains collided in New England when both train guards had different, but technically accurate, times on their watches. Thirteen passengers were killed and scores more injured.

The near instant communication of telegraphs only made the problem more confusing. Bankers in New York consulted schedules for banks in Pittsburgh while corporate headquarters for large railroads grew awash in time schedules as their lines expanded westward. One of the most dramatic illustrations on record occurred when the two ends of the transcontinental railroad were joined at Promontory Point, Utah, in 1869. Leland Stanford, cofounder of the Central Pacific Railroad, was supposed to pound in the last spike, which was wired to send a telegraph signal to both coasts. At the very least, it was a neat publicity trick. However, Stanford missed the spike, and a nearby telegraph operator keyed in a single word "Done." Though the announcement was less dramatic, it got the point across from coast to coast. America cheered the accomplishment with papers listing dozens of exact times for the historic event, all of them technically accurate.

Public debates over time keeping were common and usually centered on instituting a standardized time. In some instances, the debate took on overtly religious overtones. Does man or the heavens set the proper time? Standardized time, the argument went, was an attempt to supplant the divine creator's own watch. It was, they argued, in defiance of God's plan and an act of sinful hubris. The railroads were having none of it and finally settled the matter, somewhat uneasily, in the 1880s by instituting standardized times on their own with the government reluctantly and somewhat timidly following suit years later.

BY THE 1870S WESTERN UNION began selling time. A "time ball" rigged to the Naval Observatory in Washington by telegraph on top of its New York headquarters dropped once a day at noon for citizens to set their watches. Watchmakers, factories, and businesses could subscribe to a service that offered special Western Union battery-powered clocks in factories and other businesses linked directly to the National Observatory via telegraph. Two small Leclanché batteries in glass jars that were changed out yearly powered the clocks. The fees ran up to $375 a year for the service.

IT WAS INTO THIS TECHNOLOGICAL expansion and optimism that Cyrus W. Field stumbled. Born in Stockbridge, Massachusetts, he was a self-made man who started out as a dry-goods clerk and ended up wealthy enough to retire in his early thirties. Still ambitious and anxious for a new challenge, Field found what he was looking for in a ruined telegraph promoter, F. N. Gisborne, who had recently lost a fortune trying to run a line from Newfoundland to New York. The enterprise ended in disaster, primarily because of the rugged terrain. Not only did the line not work, but Gisborne had been hung out to dry by investors. He was arrested and his assets seized.

Still, he was optimistic and somehow managed to inspire Field to get into the electromagnetic telegraphy business. What's more, he would do it in a big way. Field's plan was much grander than simply a line from Newfoundland to New York. He would build a line from Newfoundland to Europe, approximately 1,700 miles. According to one often-repeated account, he got the idea by looking at a globe in his study just after Gisborne left his Gramercy Park town house. "It was a very pretty plan on paper; God knows that none of us were aware of what we had undertaken," Field later wrote.

In the mind-set of 1854, it was very much like planning a trip to Venus. What's more, it didn't seem to matter to Field that he knew next to nothing about telegraphy, electricity, or oceanography. Gisborne, though a ruined man, was an engineer. And Field could leverage his past successes to raise the funds. Perhaps he saw himself as a new type of industrialist for a new age. Vast fortunes had been made in ships and railroads—why not telegraphy?

Although some admired Field's oversized telegraphic ambitions, many more viewed a transatlantic cable as just another rich man's folly. As soon as word leaked out what Field was planning, the skeptics leapt on it with both feet, providing a litany of reasons why the venture was destined for failure. Hostile fish, icebergs, unknown underwater terrain, ship anchors, tides, and the simple enormity of the project were all dutifully listed as reasons for the inevitable failure of the endeavor. The British Astronomer Royal, Sir George Biddell Airy, confidently pronounced the enterprise doomed because pressure at such great depths would squeeze the electricity from the cable. Even Thoreau sourly offered an opinion in his 1854 book *Walden*: "We are eager to tunnel under the Atlantic and ring the Old World to the new, but, the first news that we will hear is that Princess Adelaide has the whooping cough."

Of course, there were also some optimists who viewed the undertaking as the final piece of technology that would bring about utopian global peace for all time.

Through it all, Field maintained an unwavering belief in the project's eventual success. In fact, he had every reason for optimism. By 1854, the technology was proven and more or less perfected. Infrastructure had arisen in the past decade to supply most of the basic equipment needed. The cable required, for instance, could be manufactured on the same machinery that produced cables—similar to rope—for mining operations and the new increasingly popular suspension bridges. And, as he would eventually discover, the underwater terrain where the proposed cable would rest off Newfoundland was particularly well suited, forming a gentle ledge to carry it eastward to Ireland.

There was also plenty of small-scale precedent. Morse, by then hailed as the inventor of the telegraph, had experimented with underwater cables as far back as 1842 with a hemp and India-rubber-coated cable. In England, Wheatstone had performed experiments in the Bay of Swansea. And, as far back as 1811, Jacobi, of failed electric boat fame, had used a wire insulated with rubber strung beneath the Neva in St. Petersburg to set off a mine. Cables had been laid across riverbeds and lakes, and two years earlier, in 1852, a cable had joined England and France across the English Channel.

To capture some credibility, Field solicited Morse to join the enterprise. How could the inventor of the telegraph resist the greatest telegraphy project in history? However, neither could he resist his old methodologies as project manager. Before signing on, he consulted with Matthew Fontaine Maury at the Naval Observatory. An oceanographer who studied currents and conducted soundings off Newfoundland, Maury discovered the ledge between Newfoundland and Ireland that made it perfect for the cable. Morse also consulted with Faraday, addressing him as a fellow scientist and seeking advice, much as he had with Henry years prior.

The tireless Field floated the stock offering, putting up a quarter of the capital himself. The venture, called the Atlantic Telegraph Company, came to life in 1856. Among the first investors were William Makepeace Thackeray, a contemporary of Dickens and the author of *Vanity Fair,* and Lady Byron, the widow of the poet, Lord Byron, and the mother of Augusta Ada Byron.

Unfortunately for Field, even the abundant enthusiasm of blissful ignorance has its drawbacks, particularly in those things technical. Oddly, the single largest obstacle the enterprise would encounter was not two miles beneath the Atlantic, but in London. His name was Dr. Edward Orange Wildman Whitehouse, and Field hired him as engineer and chief electrician.

A retired physician, amateur telegrapher, and gentleman scientist, Whitehouse was a plainspoken, commonsense man of the very worst variety. That is to say, he was arrogant, unable to admit mistakes, and conducted himself as something of a bully when challenged. Even worse, both his science and his engineering skills left much to be desired, so there was quite a bit to challenge. In the end, he turned out to be very much the villain in the enterprise.

Whitehouse made mistake after mistake with the design of the system. Some of these early mistakes were so basic they should have tipped off Field and the other investors to Whitehouse's incompetence. For instance, the cable was ordered from two different factories—Glass, Elliot Company of Greenwich and Messrs. R. S. Newall Company of Birkenhead—with somewhat vague technical specifications and a careful eye on the price.

It wasn't until a good portion of the cable had been produced that the first of many mistakes was discovered. One of the manufacturers had given the cable a clockwise orientation in turning the strands of copper while the other manufacturer had the windings running counterclockwise. The two halves of the cable were nearly impossible to splice together in any traditional manner. The anticipated stress would unwind them, much like turning the lid on a jar. The solution came by way of a complex, specially designed clamp.

The cable itself was made up of seven strands of thin copper wire, sheathed in gutta-percha, an early plastic derived from the resin of the Isonandra gutta tree native to Malaysia. It was originally imported by a Scottish surveyor, Dr. William Montgomerie of the East India Company, who hoped it might have use in surgical instruments. The venture proved less than a sterling success, but Faraday, Wheatstone, and others adopted the substance as wire insulation.

On top of the gutta-percha was a kind of tarred yarn, and finally a winding of iron wire. The cable, which measured about half an inch thick, was light, flexible, weighed about 107 pounds per nautical mile, and proved entirely unsuitable. Whitehouse's experiments, conducted on a small scale, didn't take into account the massive size of the enterprise. The diameter was far too small to carry a transmission the distance it needed to travel without benefit of a relay system. Added to the cable's woes was the quality of the copper. Either Whitehouse's vague specifications or the manufacturers' lack of quality control produced a less than ideal conductor.

When the system's flaws were pointed out to Whitehouse, he conducted a few haphazard experiments on his own, then blithely waved off the objections. "No adequate advantage would be gained by any considerable increase in the size of the wire," he wrote. After all, common sense dictated that electricity itself was "small" and there was plenty of room for a lot of it within the half-inch of tarred yarn and gutta-percha–sheathed wire. Whitehouse pushed on, ever confident he would prevail over the theoretical scientists.

Batteries on the boats carrying the cable ran constant tests. These were specially built Daniell cells filled partially with sawdust to prevent the acid solution from sloshing over the sides with the ships' rocking. However, the battery Whitehouse designed for the stations in Ireland and Newfoundland was massive. He called it the "Whitehouse Laminated" or "Perpetual Maintenance Battery." Made up of a wooden trough, the slotted cabinet held 10 pairs of platinum-coated silver plates and 10 pairs of zinc plates that totaled some 2,000 square inches of surface area. In the end, he used several of these batteries, equaling more than 300 Daniell cells. A complex mechanism allowed the operator to increase or decrease the power by lifting the plates from the acidic solution. The batteries' power was boosted by five-foot-high induction coils that jumped the current up to a frightening degree. Although accurate ways of measuring current didn't exist, some estimates place the output at upward of 2,000 volts. On the upside, as Whitehouse bragged, the cost of operation was about a shilling a day.

Three well-documented failures prevented the joining of the cable. The first attempt ended in complete disaster with both ends lost to the ocean. The fourth try proved the charm, and on August 5, 1858, the two continents were joined. Queen Victoria and President James Buchanan exchanged messages. And the entire world seemed to go "cable crazy." As with the invention of the Chappe brothers' telegraph system, pundits joyously predicted the outbreak of world peace. Surely such a miraculous device would cause even the most belligerent of the world's citizens to "make muskets into candlesticks." A story in the London *Times* enthused, "The Atlantic is dried up, and we become in reality as well as in wish one country. The Atlantic Telegraph has half undone the Declaration of 1776, and has gone far to make us once again, in spite of ourselves, one people."

*TRANSATLANTIC CABLE CROSS SECTION*

Field, who had more than twenty miles of the flawed cable left over, sold it off at a tidy profit to manufacturers and retailers who quickly set about fashioning it into trinkets. Within weeks bits of cable were turned into earrings, umbrella handles, snuffboxes, candlesticks, and some very odd commemorative displays mounted on wooden pedestals. Even Tiffany & Co. got into the act, selling lengths of the stuff at fifty cents apiece. In a newspaper story, Tiffany proudly announced, "In order to place it within the reach of all classes, and that every family in the United States may possess a specimen of this wonderful mechanical [*sic*] curiosity they propose to cut the cable into pieces of four inches in length and mount them neatly in brass ferules."

However, even before the celebrations ended, the signal began to fade. Whitehouse, of course, was confident of the solution and began pumping more current through the line by way of his large batteries and massive induction coils. After all, common sense dictated that if the signal was weak, more power was needed to span the 1,700 miles from Newfoundland to Ireland. As the signal grew increasingly faint, Whitehouse stepped up the current even more. Finally, less than a month after going into operation, the signal dropped out completely. In the end, the massive number of volts Whitehouse pumped into the cable more than likely burned out the insulation.

As expected, the press, which had hailed the technological marvel just a few weeks prior, wasted no time in declaring the entire enterprise an expensive folly. Some went so far as to label it a stock swindle and hoax.

MADDENINGLY, THE ANSWER TO VIRTUALLY every problem was at hand in the form of William Thomson (later Lord Kelvin), one of the leading physicists of the age, the best-known authority in the field of electrical science, and a member of the company's board. The son of a Scots-Irish farmer, Thomson showed early promise that kept on delivering throughout his entire life.

What Thomson told Whitehouse, who ignored the advice, was that the cable's effectiveness in carrying an electrical charge could be calculated with a simple mathematical formula based on Fourier math. So easy that a bright junior high school student could work it, the formula was based on the law of squares. Simply stated, it more or less accurately calculated that the drop-off in the current of a line is proportional to the square of the distance traveled. That is to say, a cable of two miles would have about four times the drop-off in power as a cable one mile long. So the result at the end of the line would be only one-quarter strength.

Thomson brought more than engineering math to the project. A few years previously, the German physicist and chemist Johann Christian Poggendorff had developed a new type of galvanometer for testing electrical current. More sensitive than those in use, it relied on a mirror suspended by a silk thread with tiny magnets glued to its back and a lantern's light reflected on its front. When a coil encasing the chamber where the mirror hung received the slightest charge, it would turn the mirror and reflected beam of light. The more current going through the coil, the more the mirror's magnets would react to the electromagnetic field and the more the beam of light would shift its position on a screen.

The ingenious device was capable of detecting and reacting to even the smallest current flowing through the coil. A much-updated version of the mirror galvanometer, as it was called, exists today in theatrical lighting. Shifting mirrors control lasers and high-intensity lights in theaters and nightclubs.

What Thomson had done was refine the device, making it suitable as a telegraph receiver. In fact, he had used it to test the cable on board one of the ships during an attempt to join Newfoundland and

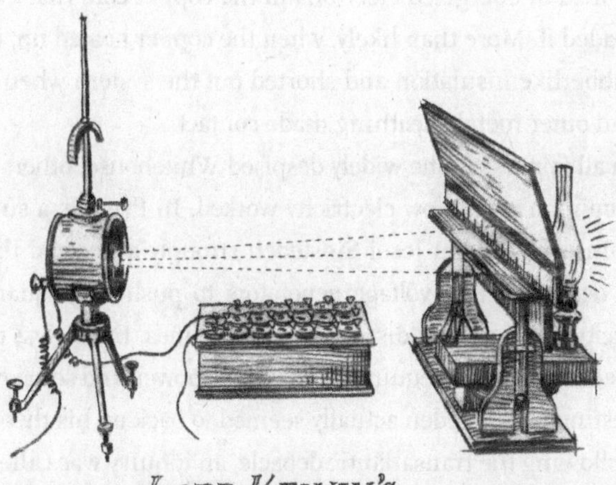

LORD KELVIN'S MIRROR GALVANOMETER

Ireland, and the device had worked perfectly. Whitehouse, of course, rejected the mirror galvanometer, though rumors persist that, out of desperation, he used it at the Newfoundland station as the signals began to fade and then swore the staff to the strictest secrecy.

Whitehouse's problem, aside from an unpleasant variety of stubbornness, was his understanding of electricity. He had imagined the electrical substance flowing from his large batteries and induction coils into the cable much as water travels through pipes. The way he saw it, electricity flowed out of batteries into the copper of the wiring. In Thomson's view, batteries served to energize an invisible field throughout the length of a wire. Faraday had also quite correctly theorized such a field. This wasn't a theory that tough-minded, commonsense men like Whitehouse, or even Field, could easily grasp. For them, it was much more likely that electricity—whatever it happened to be—was poured into one end of a wire from a battery and exited from the opposite end into an electromagnetic device.

Whitehouse's huge batteries and induction coils probably created a field of energized electrons in the copper core that eventually overloaded it. More than likely, when the copper heated up, it melted the rubberlike insulation and shorted out the system when the copper and outer metal sheathing made contact.

In all fairness to the widely despised Whitehouse, others held the same opinion about how electricity worked. In Prussia, a surgeon by the name of Wilhelm Josef Sinsteden strongly advocated the use of newly designed high voltage generators to push large quantities of electricity through long-distance telegraph lines. In an 1854 paper, he outlined his plan that required high voltage power, and some rudimentary testing by Sinsteden actually seemed to back up his theories.

Following the transatlantic debacle, an inquiry was called to look into the matter. An eight-member committee was formed, with four members from the British government, which had partially funded the project, and four from the Atlantic Telegraph Company. Witnesses were called and testimony heard. Thomson and other scientists gave lucid testimony, explaining the theories and likely reasons for the failure. In the end, Whitehouse, along with his commonsense views on electrical transmission, was shown the door. Any reasonable man would have faded tactfully into obscurity. He published a self-serving book, wrote letters, and gave interviews placing the blame on everyone but himself.

With Whitehouse out of the picture, the project was revitalized when Thomson's theories were proven in other parts of the world with other submarine cables. New money was raised and a second attempt was in the planning stages even as the Civil War raged. The deathblow finally fell on what little remained of Whitehouse's credibility when a new cable was stretched across the Atlantic. The redesigned cable, completed just months after the war's end—with two sections turned in the same direction—differed significantly from the first. The copper core was larger and of better quality, and so was the insulation. The cable also boasted a significantly wider diameter and weighed in at about three times the 107 pounds per mile of the original cable. After several mishaps, the cable was finally connected between Ireland and Newfoundland in 1866. Using Thomson's mirror galvanometer, the telegraph functioned perfectly with a modest power supply of 12 Daniell cells for a total of an estimated 12 volts.

In the end, Thomson could not resist a practical, if not dramatic, demonstration of his theories. Filling a thimble with sulfuric acid and lowering in two metals, he created a tiny battery. He then connected his battery to the cable and sent a small burst of electricity across the Atlantic that was picked up by his mirror galvanometer in North America.

JULES VERNE IN HIS CLASSIC science fiction tale *20,000 Leagues under the Sea*, published in 1870, paid tribute to the technology with a tour of the cable aboard Captain Nemo's sub.

> I did not expect to find the electric cable in its primitive state, such as it was on leaving the manufactory. The long serpent, covered with the remains of shells, bristling with foraminiferae, was encrusted with a strong coating which served as a protection against all boring molluscs. It lay quietly sheltered from the motions of the sea, and under a favourable pressure for the transmission of the electric spark which passes from Europe to America in .32 of a second. Doubtless this cable will last for a great length of time, for they find that the gutta-percha covering is improved by the sea-water.

What became clear in the wake of the transatlantic debacle and subsequent inquiry was the fact that no accurate measurement for electricity existed, at least not one easily understood by engineers. Even as the causes of the misadventure were examined, the board and witnesses struggled for language to describe exactly what had happened. Just how much electricity had Whitehouse's batteries and induction coils pumped into the cable? A lot? Too much? Far too much? A whole bunch? Even today various accounts of the failed cable provide starkly different estimates. By one widely accepted account, Whitehouse's batteries and induction coils were said to have pumped out 2,000 volts, while another estimates the number at 500 volts. The imprecision of the language was as objectionable to engineers as it was to those men of finance and industry who would potentially pay for future projects.

Scientists had been struggling for a standardized measurement of electricity for years. Describing the same concepts with different words was confusing enough, but with electricity entering the commercial realm, the situation was becoming intolerable. Engineers needed a precise way of describing and calculating the different qualities and quantities of the "subtle fluid." Committees were formed, and scientists commissioned to look into the matter. After years of haggling and no small amount of backroom politicking between the French and British, the standard units of watt, ampere, and volt emerged.

The term "volt," after Alessandro Volta, the Italian inventor of the battery, was pushed hard by the French in large part because of his support of Napoleon. Watt, for James Watt, who perfected the steam engine for industrial use, had nothing to do with electricity at all. However, he had coined the idea of horsepower as a unit of measurement, primarily as a way to make his engine's power understandable to potential buyers accustomed to equine-powered machinery. What would come to be known as the amp or ampere, was named after André-Marie Ampère, the French mathematician turned physicist who studied electromagnetic fields.

Although the transatlantic telegraph met with failure, both sides in the American Civil War understood the value of the technology on a smaller scale. Union and Confederate troops strung up more than 15,000 miles of wire and deployed mobile telegraph stations pulled by horses or mules. The impact of the telegraph in warfare was immediate and not altogether pleasant. The same kind of efficiencies delivered by improved communications to railroads and other businesses were now delivered to the battlefield with devastating results. Generals could receive reports or coordinate troop movements with more precision and speed. Plans could change quickly as new intelligence arrived from spies in the field.

General Ulysses S. Grant proved particularly adept at using the telegraph to direct troop movements with remarkable precision, while Confederate General J. E. B. Stuart hired his own wiretapper—one J. Thompson Quarles—who rode with him. Both sides tapped telegraph lines that were quickly replacing couriers on horseback. Done correctly, it was a nearly risk-free form of spying.

President Abraham Lincoln embraced the new technology with gusto, spending hours in the War Department's telegraph room, keeping in near real-time contact with his generals on the battlefield. In all, Lincoln sent nearly a thousand telegrams during his presidency, many of them in the same conversational tone he used in letters. At one point, telegraphing Grant, he urged, "Hold on with a bull-dog grip, and chew and choke, as much as possible."

IT IS DIFFICULT TO OVERESTIMATE the speed at which telegraph networks spread across the landscape. Within thirty years of Morse's demonstration in 1844, there were some 650,000 miles of cable and 30,000 miles of submarine cable linking more than 20,000 towns and villages. By 1880 there were an estimated 100,000 miles of undersea wiring connecting continents. The world was becoming smaller.

With the success of the telegraph, electricity was becoming a part of everyday life. Technology that was not readily understood by the "man in the street" was slowly integrating itself into his landscape. Predictably, odd and often outlandish theories about how the telegraph actually worked emerged as the lines continued to extend their reach, touching more and more lives in distant towns and hamlets.

Then as now, journalists delighted in retelling stories of the less sophisticated who either mistrusted or misunderstood such an obvious instrument of progress and modernity. According to press reports at the time, some simple country folk believed the wires were hollow and transported tightly wrapped written messages sent on a burst of air or acted as speaking tubes capable of carrying a voice over long distances. And in one popular story, the mother of a soldier arrived at the telegraph office with a plate of food to be telegraphed to her son fighting in the war between Prussia and France.

Today, with technology becoming increasingly sophisticated, it is not unreasonable to estimate that the percentage of the general population that understands the workings of their digital cameras or cell phones is about equal to those who understood the telegraph. However, unlike the nineteenth-century citizenry, today's population has little or no expectation of understanding the principles behind those technologically sophisticated gadgets. The devices with their intuitive user interfaces work, at least most of the time, and that is enough.

7

# *Finally, Something Useful*

> ***"Electricity is but a new agent for the arts and manufactures, and, doubtless, generations unborn will regard with interest this century, in which it has been first applied to the wants of mankind."***
>
> —*Alfred Smee,*
> Elements of Electro-Metallurgy, *1852*

It didn't take long for clever inventors to start scaling down the general principles of the telegraph for consumer products. Electric doorbells became the rage among those who could afford them, while burglar alarms and police call boxes also began to emerge. The first electronic call boxes for fires were installed in Boston in 1852, replacing a system that depended on ringing church bells. Mechanical bell pulls and speaker tubes used to summon staff in some luxury hotels were replaced with simple telegraphic systems comprised of little more than a button and a bell. These "annunciators" often had their central control board set up in a prominent position in the lobby near the main desk, putting it on display as a working symbol of modernity and efficiency.

One Victorian era inventor of a morbid disposition capitalized on the then common fear of being buried alive. Devices that could be activated from inside a buried coffin, such as a line to pull and ring a bell, breathing tubes, and other crude methods of communication from the grave already existed on the market, but this wily inventor brought out a state-of-the-art device that is best described as the first electric doorbell for a coffin. The prematurely interred simply pressed a conveniently located button to ring an electric bell on the surface. In another version of the device, movement inside the coffin would activate the alarm. And one inventor in Topeka, Kansas, took out a patent as late as 1891 for an electrical device that attached to the deceased hand, making signaling convenient and easy.

The clicking and printed binary code of dots and dashes in the nineteenth century held as much appeal as the hidden binary code of computers' machine language of 1's and 0's in the late twentieth century. In the rapidly expanding newspaper trade, periodicals quite literally adopted the name "telegraph" to denote speed and breadth of coverage. News from foreign countries that had once taken weeks and even months to reach New York, London, or Paris was now transmitted within days or even hours. Early on, Paul Julius Reuter, the German journalist and friend of the scientist Carl Friedrich Gauss who would go on to establish the global news agency, traded in his carrier pigeons for a telegraph system to transmit reports from the London Stock Exchange.

Dubbed "the highway of thought," the emergent telegraph industry attracted the most ambitious and brightest of young men to its ranks. Thomas Edison was a telegraph operator in his early years. The filmmaker Alfred Hitchcock worked for a company that supplied equipment called the Henley Telegraph Company.

> Well, little boys are always asked what they want to be when they grow up, and it must be said to my credit that I never wanted to be a policeman. When I said I'd like to become an engineer, my parents took me seriously and they sent me to a specialized school, the School of Engineering and Navigation, where I studied mechanics, electricity, acoustics, and navigation . . .

Working for the telegraph company while still a teen, Hitchcock started out as a "technical estimator" for cables before transferring to the advertising department as an illustrator. Alfred Vail's younger first cousin, Theodore, started his career as a telegraph operator and later proved instrumental in the formation of AT&T as a monopoly.

Telegraphy was seen as a solid career with a bright future for a bright young man. Far from the Industrial Revolution's mills and mines with their dirt and danger, telegraph operators were in the vanguard of a growing middle class. In his later years, steel magnate Andrew Carnegie fondly recalled his early days as a telegraph messenger, writing in his autobiography, "I do not know a situation in which a boy is more apt to attract attention, which is all a really clever boy requires in order to rise."

The impact of the new technology on metropolitan centers was very rapid indeed. In London, telegraph lines multiplied exponentially until the traffic at the headquarters became so heavy operators couldn't keep up with transmissions to the stock exchange. Timely information provided by the telegraph, over even relatively short distances, evolved quickly from a novelty to convenience and, finally, into a competitive necessity in the financial capital.

At least part of the London problem of overloaded lines and operator backlogs was solved by Josiah Latimer Clark, one of those scientific and engineering polymaths the nineteenth-century British Empire seemed to manufacture in surprising quantity, almost as if turning them out on an assembly line. Trained in chemistry, Clark switched to civil engineering on the railroads, then electricity. Later, he would develop his standard cell, which produced a little over one volt, to calibrate instruments. He had also taken an active interest in astronomy. Whatever field he touched, no matter how briefly, he seemed to invent new devices for it, and he even sat on the board of inquiry of the first Atlantic cable.

To solve the problem of overloaded short-distance lines, Clark came up with a system of pneumatic tubes. Just as the telegraph was being scaled down to power doorbells, Clark scaled down a failed concept for a pneumatic tube rail line that dated back to 1810 and had originally been intended as a type of high-speed people mover. In 1853, he installed the first line of tubes measuring an inch and a half and spanning some 200 yards from the central telegraph office to the stock exchange.

Telegraphers would transcribe the messages coming through the wire and then shunt them off in felt bags in the tubes to the stock exchange. A six-horsepower steam engine powered the whole operation. The tubes proved so efficient that they were soon adopted for use in ever-expanding office buildings and the relatively new concept of department stores, where they would remain a staple of efficient short-range communication well into the twentieth century. From the inadequacy of one technology arose another less sophisticated technology to take up the slack.

VERY EARLY ON IT BECAME apparent that the speed information traveled was particularly essential to the financial industries. Speed provided not only a competitive advantage, but also new opportunities. News from far-flung corners of the United States began flowing into New York, and with the opening of the transatlantic cable, a lucrative business in arbitrage sprang up with stocks traded on both the London and New York exchanges. To almost nobody's surprise, unscrupulous speculators soon saw potential in the rapidity of the telegraph to spread rumors that started runs on commodities and stocks they could then quickly sell at a profit. As much as we often like to believe the past was somehow more genteel than our current age, chicanery in the financial markets did not arrive with the twentieth century.

Despite the long-distance advancements, a small army of messengers was still employed at every major stock exchange to carry messages of price changes from the trading floors to the brokerage houses. Called "pad shovers," these messengers gathered on the floor of the exchanges and wrote down the quotations of a single stock or small group of stocks, which they would then deliver to the brokers. By any measure, it was a hugely inefficient method, with each brokerage house employing a dozen or more young messengers prized for their speed and accuracy. Time really was money, and a quick pad shover could boost profits in the hurly-burly of Wall Street. The old saying that "Wall Street starts in a graveyard and ends in a river" was never more true than in the 1860s. Fortunes could be made or lost overnight in that bare-knuckled world of brokers, financiers, speculators, and investors.

TO FINANCE THE WAR EFFORT, President Lincoln suspended the gold standard and issued some $450 million in paper money, which were soon dubbed "greenbacks." However, gold was still used for international trade and tariffs, among other things, and with the war's outcome far from certain, was more trusted than Lincoln's paper money. Without the backing of gold, the value of the new currency fluctuated wildly, falling against gold with every Confederate victory and rising with each Union success on the battlefield. Anyone with insider knowledge of the conflict could make a fortune speculating in gold, no matter which side seemed to be winning the war.

At one point, the banker Jay Cooke, showing patriotic outrage, called the gold speculators "General Lee's left flank." To speculate in gold was widely viewed as unpatriotic, and eventually the New York Stock Exchange banned gold trading altogether. The gold traders, apparently unfazed by the banishment, simply packed up and moved around the corner to William Street, where they opened the Gold Exchange and went on about their business.

Shut down briefly in 1864 after gold peaked at $300 in greenbacks per $100 in gold, the exchange was reopened a short time later. With the end of the war, prices calmed down quite a bit, while the exchange remained a hectic financial bazaar of speculators, investors, and a swarm of pad shovers dispatched to write down the latest gold prices and hurry them back to merchants and brokers.

Samuel Spahr Laws arrived at the Gold Exchange straight from Paris. Born and raised in West Virginia, he had graduated from Princeton Theological Seminary, where he studied under Joseph Henry, and had presided over a Presbyterian congregation in Missouri. Then in 1861, refused to sign an oath of allegiance to the federal government, an act that earned him a year in prison, which he spent reading philosophy. After his release, he headed off to Europe, returning to the United States in 1863 and settling in New York. Despite his somewhat eccentric résumé, Laws seemed to impress his bosses, rose quickly, and within a year was vice president of the Gold Exchange.

An amateur electrician and tinkerer, no doubt inspired by Henry from his days in Princeton, he came up with a device to ease the crush of messengers on the exchange floor. Called the Gold Indicator, the device displayed current gold prices on a two-faced mechanism of rotating drums. Like a double-sided clock face, one side offered the latest prices to traders on the exchange's floor, while the other side faced out on the street for the messengers. Patented in 1867, the battery-powered display was simply a short-range telegraphic network, with its line extending no farther than the trading floor of the exchange.

It was not long afterward that Laws hit on the kind of idea that makes men rich. Why stop at just a single indicator? Again, Metcalfe's Law kicked in: if one Gold Indicator was good, then hundreds were even better, and potentially profitable. Laws quit his position on the exchange and formed the Gold Indicator Company to provide the devices on a subscription basis, linked by telegraph wires from a central exchange to any business that needed or wanted one. The invention, combined with a printing function, was a small success.

By 1870, Laws sold the company and became modestly wealthy. Possessing a restless mind, he received a law degree from Columbia University and was admitted to the New York bar, then quickly earned his medical degree at Bellevue Hospital by 1875. Still restless, he eventually left New York to become president of Missouri State University (now the University of Missouri).

However, Laws's greatest contribution might have been to help a young man in need. Down on his luck, the young itinerant telegrapher Thomas Alva Edison arrived in New York in 1869 nearly broke after a failed series of business ventures. Although a skilled telegraph operator, Edison was having some difficulty finding work, and a friend let him sleep in the battery room of the Gold Indicator Company.

It was at this low point in his career that luck seemed to find Edison. When the Gold Indicator system broke down, the company's office was quickly mobbed with messengers seeking answers. With the price of gold fluctuating rapidly and the traders now dependent on the indicators, a delay of even an hour or two could mean a lost fortune. "Within two minutes over three hundred boys crowded the office, that hardly had room for one hundred," Edison later recalled, "all yelling that such and such broker's wire was out of order and to fix it at once. It was pandemonium."

The problem, as Edison quickly saw it, was that a spring had become disengaged and fallen between gears to jam a critical mechanism. "Fix it! Fix it! Be quick!" Laws demanded. Edison fixed it and was soon rewarded with a job at $300 a month, more than enough to get the young inventor back on his feet and into a position that would provide access to some of New York's richest moneymen.

Although the technology was not overly complex, Laws's device was revolutionary for a few reasons. The central station and its employees did all of the technical work, essentially laboring in an early version of the data processing center. Here was a complex electrical communications device that didn't need a full-time operator at the customer's terminal. Brokers could place the oblong indicator box in their office and follow the price of gold as it clickety-clacked happily away without a technician to maintain the machinery or understand the messages. The imperfect analogy would be the introduction of graphical interface on desktop computers that eliminated the need for expertise of typing in code. Laws's Gold Indicator also provided a steady flow of information in near real time at a distance.

Edward A. Calahan took up the concept in 1867 and ran with it. A telegraph operator by profession, he had worked as a Wall Street pad shover as a boy and he came up with the idea of the stock indicator. It was essentially a telegraph system comprised of a clockwork-type mechanism with two synchronized printing wheels—one for the price and another for the company name—under a bell jar.

Others had designed similar systems, but Calahan was the first to bring a fully functional model to marketing in a big way. Shortly before Christmas in 1867, Calahan installed the first ticker in the brokerage house of David Groesbeck and Company. Brokers gathered around the machine in the early morning, then cheered as the first prices came through, transmitted by Calahan's agents on the floor of the exchange. Because of the clattering sound its printing components made, the system was very quickly dubbed the "ticker."

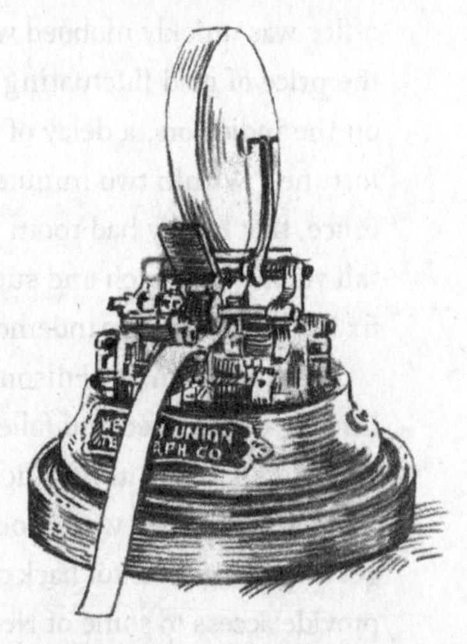

*EARLY STOCK TICKER*

Calahan quickly signed up hundreds of subscribers for the fee of $6.00 a week. Still, his first model wasn't perfect. The wheels frequently fell out of alignment, causing the stock price on the bottom and the name of the company on the top to blur together on the narrow strip of paper. The batteries also proved an unexpected problem. Each ticker included its own zinc and copper battery, usually placed on the floor beneath the unit with a pair of uninsulated wires running up to the little electric motor. Messenger boys who previously delivered quotes were now pressed into servicing the batteries, making twice-weekly deliveries of acid before the exchange opened. The sloshing sulfuric acid soon became notorious for ruining the fine carpeting of many of the leading brokerage offices, along with the pricey clothing of brokers. Calahan quickly solved the problem by providing power from a central station.

Not everyone was so enamored with the ticker. A few "old timers" set in their ways insisted on keeping the messenger boys on the payroll, just in case, and at least one broker, by the name of Bill Heath, nicknamed "The American Deer," continued to make his rounds, running from the exchange back to the brokerage house to shout out the latest quotes above the clatter of the tickers.

To power the growing number of devices, battery technology was now emerging as an industry rather than just a tool for the lab. The old-style batteries, such as Grove's, Wollaston's, and even Cruickshank's, were better suited for lab experiments or telegraph offices than for use in the field. Relatively expensive to build, they also required near-constant attention. In telegraph offices, for instance, the batteries used were necessarily large and required regular care by an employee known as the battery man who would tend to their maintenance in much the same way a train's firemen tended to the steam engine's boilers. This was painstaking work performed on strict schedules.

Handbooks of the era are filled with pages of detailed instructions for the general care of the battery. In one of the most popular, *Modern Practice of the Electric Telegraph* by Frank L. Pope, published in the 1880s, the author sets out specific, often exhaustive, guidelines for battery maintenance. For the "renewal" of a common Daniell cell, he offers the following instructions:

> In renewing this battery the zincs should be scraped and well cleaned with a stiff brush, the porous cups thoroughly washed, and the old solution contained in them thrown out, with the exception of about one third of the clear portion, which should be returned, otherwise the battery will require some hours to recover its full strength. The copper deposit upon the zincs is valuable, and should be preserved.
>
> Every two or three months the coppers ought to be taken out and the deposit upon their surface removed, which may be done two or three times. When they become too much encrusted to afford room for the porous cups they must be replaced by new ones.
>
> Porous cups ought to be renewed whenever they become too much encrusted with copper. If cracked they should be changed at once, otherwise a great waste of material will ensue.
>
> The crystals which form around the edge of the outer jar require to be occasionally wiped off with a damp cloth, or they will eventually run down the outside and form a connection between the jars, giving rise to a great consumption of material without corresponding benefit.

The nineteenth century saw the battery as a specialized industrial tool—not all that different from the way 1960s computer professionals saw the unwieldy room-sized systems that noisily sorted punchcards and held data on magnetic tape.

8

# Power and Light

*"The Electric Girl Lighting Company will furnish a beautiful girl of fifty or one hundred candle power, who will be on duty from dusk till midnight—or as much later as may be desired . . ."*

*—New York Times*

In the nineteenth century, as today, the world was in need of a better battery and there was no shortage of inventors willing to try their hand. Edison began experimenting with batteries during his free time very early on. Although his attempts to invent a battery that wasn't subject to the dreaded polarization or the diffusion that seemed to accompany most antipolarizing techniques of the day have been largely forgotten, he put in a significant amount of time and effort on the search. After hitting on a partial solution early in his career as a telegrapher, he quickly dashed off a letter to Latimer Clark regarding his experiments. And, according to one often-repeated account, while in Louisville, Kentucky, he lost his job at the local telegraph office when sulfuric acid for an experimental battery he was working on spilled and leaked through the floor and into the manager's office below.

The catalogs of telegraphic and lab equipment companies as well as popular books on electricity are stuffed with descriptions of batteries with exotic-sounding names: Smee's battery—named after its British chemist inventor, Alfred Smee, who sought to reduce polarization by roughing up the sides of the plate, Walker's platinized carbon battery—popular among telegraph offices in England—Tyer's, Baron Ebner's, and Maynooth's battery, invented by Rev. Nicholas Callan.

The idea of a battery with an actual name—beyond familiar corporate brand names—strikes us as beyond quaint. We live in an age when batteries are largely viewed by the nontechnical general public not even in terms of specifications, such as voltage, but rather in size through convenient designations, such as AA or AAA. And too, batteries are increasingly unseen as more and more devices rely on the more convenient recharging rather than replacement.

Unlike today's batteries, nineteenth-century batteries were considered a prominent component of whatever device they powered. James Clerk Maxwell, for example, in writing his *Valentine by a Telegraph Clerk,* not only wryly compares the smitten clerk's love to a variety of specific batteries, but also to their most prominent technical attributes. They were considered as salient a part of the technology as the network's wires or key.

IN 1866 THE FRENCH ENGINEER Georges Leclanché made the next technological leap forward in battery design. Leclanché's contribution was little more than a glass jar filled with ammonium chloride (often called sal ammoniac), a positive electrode of manganese dioxide, and a negative of zinc with a small bar of carbon thrown in. It was, in almost every respect, the perfect battery for simple applications such as doorbells. It could be manufactured inexpensively and en masse, and its chemistry could be charged cheaply and easily. Millions were manufactured, and tens of thousands of them found uses in telegraph equipment and later in telephones before central exchanges provided power. Leclanché's "wet cell," as it was popularly referred to, pumped out 1.5 volts and is generally seen as the forerunner to the world's first widely used battery, the zinc carbon cell or dry cell.

The wet cell was the first battery not to use a diluted acid solution. When the battery seemed to discharge completely, users simply dumped out the old electrolyte and replaced it with fresh sal ammoniac as a corrosive. Though its history includes use in alchemy, where it was portrayed as one of the four "spirits," it was also a common household chemical used in baking, cleaning, and even the production of licorice. Easily available and relatively safe, sal ammoniac was more corrosive than Volta's brine solution, but less corrosive than acid, though still strong enough to pull electrons off metal and create a charge. Another battery, called the gravity battery or crow's foot because its electrodes hung down into the solution, was also popular.

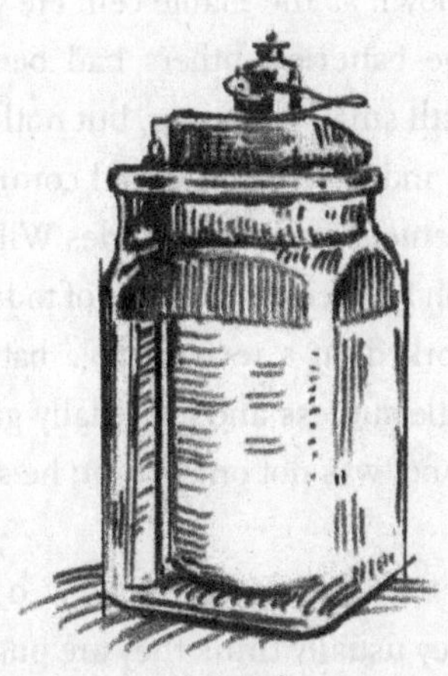

LECLANCHÉ BATTERY

The Leclanché battery did have one fairly large drawback. It was only good for intermittent use. It ran down relatively quickly but regained much of its full charge nearly as fast when allowed to sit for a while. This meant the Leclanché was virtually useless for things like electric lighting or stock tickers, any application that required a sustained flow of current. At a time when there were few electrical devices to power, this was not a particularly significant shortcoming. However, as the number of devices grew, a constant, reliable power source would be needed.

IN THE LONG HISTORY OF battery inventors, perhaps no individual showed more dogged determination than the French engineer Gaston Planté, inventor of the first practical rechargeable battery or storage battery. Planté, who died at age fifty-five in 1889, spent an astounding thirty years developing the lead acid battery that became known as the Planté cell. He was not the first to try his luck at storage batteries; others had been experimenting and even meeting with small successes, but nothing was developed that could be used in industry or produced commercially. For instance, Karl Wilhelm Siemens (later Sir Charles William Siemens), the brother of Werner von Siemens, cofounder of today's electronics giant Siemens AG, had worked on a rechargeable battery based on the Grove design with little success and eventually gave up the effort. Planté, on the other hand, was not one to quit; he spent three decades on the project.

WHEN PEOPLE THINK OF RECHARGEABLE batteries at all, they usually think they are putting electricity (whatever that is) back into the battery, very much like filling a drinking glass from a larger pitcher. In fact, when we recharge batteries, we are simply returning them to their original state. That's the difference between primary cells or standard batteries and rechargeable or secondary batteries. In the primary batteries, the changes can't be reversed because of the chemical composition and arrangement of the metals used, but in a rechargeable battery, they can be returned to their original chemical state by reversing the flow of current.

Very simply, when a battery discharges, it releases electrons from a metal and through the negative electrode to external circuitry that powers a device and then back into the positive electrode. During normal use, a battery's negative pole becomes oxidized, sending off electrons, while its positive pole gives up oxygen. When a storage battery is recharged, the positive pole is oxidized and the negative pole reduced, shedding the positively charged ions it accumulated during use.

By the mid-nineteenth century scientists knew with reasonable certainty that batteries didn't run out of electricity, but rather that something happened to the metals inside. The answer to extending a battery's life had to involve finding a way to either slow down or reverse those chemical changes. At first they experimented with various substances, such as lead peroxide, to slow down the polarization process, but this usually led to other problems in the chemistry.

Like other battery scientists of the time, Planté experimented with all manner of metals, ranging from tin and silver to gold and platinum. Eventually he settled on lead. It was readily available, inexpensive, and possessed most of the qualities needed. But very quickly Planté discovered that lead also delivered a whole new set of problems. For one thing, its surface wasn't porous enough to allow much of the acid in a battery to penetrate and release electrons. What he needed was a metal with a lot of surface area, something microscopically resembling a sponge. Lead more or less resembled silk. The answer to this problem was as simple as it was time consuming—let nature take its course. In the use of lead batteries, the surface area accumulates a layer of peroxide that is porous. So what Planté did, essentially, was polarize the plates with Grove batteries and allow them to self-discharge, then charge them again and allow for self-discharge, repeating the process over months to prematurely age the battery. It was, by all accounts, a hideously tedious process, which he named "formation."

Planté was motivated, at least in part, by the commercial potential of a high voltage storage battery capable of reliably producing a constant current. Throughout his decades-long endeavor he maintained close ties with industry and businesses, such as Breguet, which was then selling batteries along with a range of telegraph equipment. In nearly every sense, he was closer to the modern (research & development) engineer than the scholarly scientist.

And why not? A mass-produced storage battery would be a hot commodity, and not just in telegraphy. In medicine, for example, it had been known for some time that certain metals such as platinum glowed when a sufficient amount of current was run through them. Heating platinum wire could produce a light to illuminate body cavities or a hot filament to use as a cauterizing tool.

Then there was Saturn's tinder box, which Planté patented in the early 1870s. The user pressed a button on the side to send current through a platinum wire that glowed hot enough to light a cigarette or cigar while the battery inside the box also powered a doorbell. The gadget seems a simple thing, but Planté may have been far ahead of his time in combining these two unrelated functions in what was then seen as a technologically sophisticated device. The same basic principle applies—a single power source and multiple functions housed within one package—in today's cell phones, which include features such as GPS, music, and Internet access.

Another possibility was to use the storage batteries to power lighthouses. Large batteries could replace the traditional oil lamps with more powerful limelights or carbon arcs. One idea was to use Grove cells to recharge primary batteries, which provided a higher constant voltage. In effect, they would work as a kind of induction coil that modified and regulated the current, acting as load levelers. While the Grove cells would ultimately prove too expensive, by the 1870s the Gramme generator—one of the first commercial dynamos available, named after the Belgian inventor Zénobe Gramme—eventually did the trick.

PLANTÉ ALSO HARBORED VAGUE PLANS for scaled-down versions of the lighthouse scheme for indoor lighting in homes. Although still not realistic in the 1870s, electric lighting was viewed as something of an eventuality. As far back as the 1840s, scientists in Europe were toying with the idea. In 1848, for example, when Thomas Edison was barely a year old, Joseph Swan, a chemist, was already experimenting with incandescent lighting in England. He had continued his experiments, which included carbonized paper in a vacuum, off and on for more than thirty years. The roadblocks he faced were due largely to the equipment available at the time. Vacuum pumps weren't very efficient, and, too, he used thick, low-resistance carbon rods that required massive amounts of electricity to heat up, while Edison used high-resistance, very thin filaments of carbonized bamboo.

By the early 1870s Swan was already demonstrating a somewhat imperfect, but working, electric light. And, by 1879, he managed to light a street in Newcastle upon Tyne with arc lights. A detailed account of his work was published in the July issue of *Scientific American,* just a few months before Edison began experimenting with carbon. Whether Edison, who had a lifelong habit of appropriating and improving on the ideas of other inventors, read the article about Swan's lighting scheme, remains a point of debate. However, enough elements of Swan's bulb were included in Edison's final model to trigger a patent dispute in Europe that ended with a partnership in England called Edison & Swan United Electric Light Company, Ltd., later simply Ediswan.

Other inventors and scientists were also hard at work in Europe and beyond. Some of them, such as the Britons Warren De la Rue and Frederick de Moleyns, used platinum filaments in their designs. However, electric lighting dated even earlier in specialized uses, such as the theater. In 1849, for example, the Paris opera was using arc lamps powered by batteries as a special effect in some performances. This was an expensive undertaking, but it packed the house. The use of special effects became so popular that in 1855 the Paris opera hired an electrical expert, L. J. Duboscq, who incorporated as much electrical light as possible into performances, including a *laterna magica* (magic lantern) for projecting images on a screen.

Then in 1872, the Russian Aleksandr Lodygin installed more than 200 electric lamps around the Admiralty Dockyards in St. Petersburg, but he apparently used far too much current and burned out the carbon filaments within hours. But Lodygin was nothing if not enthusiastic regarding electricity's potential. He once planned an electric helicopter, which never came to pass, though, using a tungsten filament, he did perfect a lightbulb prior to Edison before going on to patent several other electrical devices, including electrical motors, electrical welding tools, and even an electric oven.

Another Russian, the engineer Pavel Nikolayevich Yablochkov (sometimes known as Paul Jablochkov), who worked in the telegraph industry, was also having some success with electric lighting. His "electric candle," essentially a small version of the carbon arc in an ornate holder, caused a sensation when it was demonstrated at the 1878 World's Fair in Paris. And then there were two Canadians, James Woodward and Mathew Evans, who came up with a working light in the 1870s.

By the time Edison began his experiments in earnest in 1878, the field had already developed a large body of knowledge. Specifically, he had two pieces of the puzzle already solved: the Yablochkov system that lit multiple "electric candles" simultaneously in a single circuit, and a powerful generator called the telemachon that lit eight bulbs at once, also on a single circuit, which he had seen at another laboratory. This was no small problem for nineteenth-century engineers—how to light multiple bulbs on a single circuit, so that if one bulb blew out or was turned off, the others would continue to function.

In a newspaper interview with the *New York Sun*, Edison described his visit to the lab enthusiastically, " . . . I saw for the first time everything in practical operation. It was all before me. I saw the thing had not so far to go but that I had a chance. I saw that what had been done had never been made practically useful. The intense light had not been subdivided so that it could be brought into private houses."

Throughout his most productive years, Edison's product development style had more to do with incremental improvements of previous inventions than original ideas. He would regularly dispatch employees to study patents and journals and then write up summaries of what they found. Once he decided on potential solutions, he would divide the work between teams of craftsman and technicians. Edison viewed even the experimentation element of his process as a business, keeping detailed records of expenses associated with each project and experiment.

Edison, who had read Faraday's *Experimental Researches in Electricity,* called the English scientist the "master experimenter." He would, eventually, adopt Faraday's position that even failed experiments were capable of yielding valuable data. Paradoxically, the very thing that held Faraday back in science—the absence of a strong mathematical foundation—would inspire and propel Edison forward in engineering.

This type of research more closely resembled modern product development than the cutting-edge work of Henry or Faraday. Although much has been written about Edison the inventor who doggedly pursued solutions through laborious trial-and-error experimentation, relatively little has been set into type about this decidedly less glamorous, though more businesslike and pragmatic approach. In many respects, Edison didn't want to reinvent the wheel, just build a better wheel, and then sell it in quantity.

In the popular media, Edison fashioned for himself an image as the humble tinkerer, the hard worker whose gift of genius did not crush his folksy ways. He spoke plainly, not burdened by an oversized ego or entranced by arcane scientific mumbo-jumbo. Far from the otherworldly absentminded professor, he carried with him all the plainspoken credibility of the common man. The Wizard of Menlo Park would leave the obscure glories of science, theories, and publication in scholarly journals to the scientists. He was a simple man, simply making products ordinary people could use and enjoy.

What most of the public didn't see was his unrelenting drive and business savvy. Edison had come of age in the nascent corporate worlds of the telegraph and the railroad. There could be no better place or time for an ambitious young man to learn the basic principles of technology and business. Western Union, in particular, promoted study and the acquisition of scientific knowledge among its employees, including telegraph operators. This was not a purely altruistic policy. With a scarcity of technical expertise, the company sought to grow its own cadre of technicians and engineers to promote up through the ranks. The company magazine, *The Telegrapher,* welcomed articles on technical innovation by employees with enough ambition to tinker with the technology in their spare time. And Edison was nothing if not ambitious to the point of obsession for most of his life. In a notebook entry dated in the early 1870s, he wrote of his first wife, "Mrs. Mary Edison my wife dearly beloved cannot invent worth a damn!"

Edison, who crafted his image as carefully as Morse, had an advantage over the inventor of the telegraph when it came to public relations. He knew newspapermen from his days as a young itinerant telegraph operator. Always skillful as a telegrapher, he rose to the top to become one of the chosen few entrusted to transmit news stories. His early experiences with reporters and as publisher of a pair of small newspapers provided invaluable insight into public relations. As reporters came to learn, Edison could be counted on for the colorful quotation and a minimum of scientific or technical jargon.

Like Morse, there was a good amount of truth mixed in with Edison's self-created media persona. His story really does fit the mold of the nineteenth-century ideal of hard work and perseverance lifting the poor boy from obscurity and poverty to fortune and fame, though what is often left out is his utter ruthlessness when it came to the business of business. At one point he collected a hefty fee from

Western Union to improve on Alexander Graham Bell's telephone in an attempt to break the young inventor's patent. The effort would have succeeded if not for Bell's father-in-law, who aggressively and successfully defended the patent. Edison was once called "the professor of duplicity," but perhaps the most devastating description came from his friend William Orton, the president of Western Union, who was reported to have said, " . . . that young man has a vacuum where his conscience ought to be."

Edison, of course, went on to create a viable electric light, patenting the bulb in the United States in 1879. Certain that electrical lighting would find its power from a central station, his eventual plan called for nothing less than the creation of an entire electrical infrastructure to supply power in much the same way gas was delivered. That meant generating stations along with power lines, metering systems, and work crews to maintain all of it.

Not unexpectedly, Edison had sound technological reasons for his stance. By the 1870s, large dynamos—as electrical generators were called—were coming into their own. Chief among these were the Zénobe Gramme dynamos, built for industrial uses, such as electroplating. Gramme, who started as a carpenter before entering the field of electricity, spent years perfecting his generator, which proved to be the first really efficient electric motor. During an 1873 exhibition in Paris, a workman accidentally wired one of the dynamos to another power source. The dynamo began spinning. It didn't take long for the Belgian engineer to realize that in the process of improving his dynamo, he had created an electrical motor that was more efficient than Faraday's philosophical toys—it could perform work.

Others in the field were not as certain as Edison when it came to the future of electric power. For one thing, would every consumer even *want* electrical power? Some imagined each home with its own minigenerating plant that ran on coal. For many, the storage battery was very much seen as the future of domestic electricity, at least in some quarters. Companies were formed in England and France with shares sold to finance production and sales.

According to this school of thought, the obvious luxury of electricity was destined to light only the homes of a fortunate few, powered by powerful storage batteries inside the home. One idea was to deliver the batteries as needed to power electric lights. This scheme, which never really got off the ground, was popularly known as the "milk bottle" plan, since the batteries would be delivered as milk was. People (or their servants) would leave their depleted batteries by the front door and new, recharged batteries would be delivered on a regular schedule.

A similar plan had two sets of storage batteries placed in the home. While one set was recharging from a central station, the other set would be in use providing a steady current. Dozens of patents were granted for electromechanical switches to run the system. Another idea that enjoyed brief—very brief—consideration was for houses to store a series of large iron tanks in the basement that were filled with an alkaline solution and proper metallic anodes and cathodes. When these enormous batteries were fully discharged, the zinc anodes oxidized and fully dissolved in the solution. Home owners would then drain the electrolyte into another iron tank and run carbon dioxide through it to produce white zinc they could sell at a profit. Not a new idea, telegraph offices had been doing it for years to offset costs, but it was still wholly impractical for consumers.

Given the times, none of these schemes was altogether outrageous. Most private homes depended on few, if any, public utilities. The majority of homes in the United States had their own wells or cisterns, coal heating systems, and outhouses. In urban areas there were gas companies that ran pipes into homes for lighting, but the infrastructure was not as well developed in the United States as in England.

Of course, Edison, who was planning his own large generating plants, was not a fan of battery lighting. He had investigated storage batteries early on as a way to store energy in power plants to provide an even flow of current that would extend bulb life, but found them wanting and eventually abandoned the concept. "The storage battery is, in my opinion, a catch penny, a sensation, a mechanism for swindling the public by stock companies," he was quoted as saying in an 1883 London interview. "The storage battery is one of those peculiar things which appeal to the imagination, and no more perfect thing could be desired by stock swindlers than that very self same thing . . . Scientifically, storage is all right, but, commercially, as absolute a failure as one can imagine."

EDISON WAS NOT THE FIRST to light a city, though he was the best marketer of city lighting. Two years before he revved up the dynamos for this first power plant on Pearl Street in Manhattan, another firm, the Brush Arc Lighting Company, installed more than twenty arc lights in the city to light a street at its own expense. It was a neat promotion, but no match for Edison. With backing from J. P. Morgan, Edison was soon in business, installing stand-alone generators to light high-profile locations, such as the New York Stock Exchange, Chicago's Academy of Music, and even shop windows. Within a few years he installed more than 300 of these generators across the country.

This didn't mean Edison was retreating from his original position on central power; rather, he was promoting lighting by the most efficient means available. In today's marketing parlance, he positioned electricity as "upmarket" and exciting. In lower Manhattan, he supplied a fashionable theater called Niblo's Garden with battery-powered bulbs that dancers wore as part of their costumes. To Edison and others working in the field, anything that put electrical lighting in the view of the general public was a good thing. The store windows and signs that Edison lit in large cities were not just selling the products of their sponsors, but the concept of electric lighting as well.

Just as the telegraph networks had been scaled down to doorbells and hotel annunciators, electric lighting, too, was scaled down. In

Europe and the United States it was possible to rent electric lighting for special events. By the 1880s catered electricity came complete with strings of electric lightbulbs and lead storage batteries to power them. Party hosts negotiated with the company as to how many lights they wanted and how long they wanted them lit. Trained electricians, schooled in the mysterious arts of the technology, tended the lighting arrangements. Feasting and dancing under the battery-powered luminescence was viewed as the height of modernity and fashion. The power was supplied via large lead acid storage batteries provided by companies such as the New York Isolated Accumulator Company and the Electric Storage Battery Company (known as EPS), which made lighting the balls of the rich something of a specialty. No doubt today such gatherings would be candlelit.

The 1880s also saw the formation of the Electric Girl Light Company based in New York City. For a fee, party hosts could rent young ladies decorated with electric lights powered by small batteries. The company's launch was described with a bit of wit in the *New York Times*. "The Electric Girl Lighting Company will furnish a beautiful girl of fifty or one hundred candle power, who will be on duty from dusk till midnight—or as much later as may be desired," the story in the *Times* enthused.

> This girl will remain seated in the hall until someone rings the front doorbell. She will then turn on her electric light, open the door, and admit the visitor and light him into the reception room. If, however, any householder should desire to keep the electric girl constantly burning and to employ another servant to answer the bell, there can be no doubt that the electric girl, posing in a picturesque attitude, will add much to the decoration of the house.

The Electric Girl Lighting Company was not the only one taking advantage of battery-powered lights. The innovative dancer Loie Fuller incorporated electric lights into her choreography, wiring her dancers with bulbs and batteries to perform on a darkened stage. Mrs. Cornelius Vanderbilt, the railroad mogul's second wife and a grand dame of high society, was known to commission gowns and dresses with electric lights with which to stage tableaus—posed still lifes—to entertain dinner guests.

For those of more modest means there was still access to electricity or at least the promise of battery-powered electric miracles for everyday use. Labor-saving household products that included everything from the first electric iron and electric fan to the sewing machine and toaster were all patented and often publicized. Never mind that many of the products were years, even decades away from practical use, they showed what was possible with the power of electricity.

The battery also found some unusual uses. It was said that the African explorer and journalist Henry Morton Stanley (of "Dr. Livingstone, I presume?" fame) carried a small battery during his 1870s African expedition that gave tribal leaders a shock when they shook hands in order to instill in them a sense of his superiority and power. When the trick received criticism, one defender wrote, "It is beyond understanding why fault should be found with this harmless and efficient method of teaching a truth."

THE 1890S WORLD'S FAIRS BECAME high-voltage showcases for electricity. The brute force of steam-powered marvels of the Industrial Revolution that had crowded the exhibitions just a few years previous were quickly giving way to the electric miracles as the twentieth century approached. At the World's Columbian Exposition of 1893 in Chicago the fairgrounds were lit day and night by more than 90,000 incandescent bulbs powered by generators in a display of more electrical lighting than any city in the country. Newfangled devices such as moving sidewalks and more than fifty battery-powered boats in the man-made lagoons thrilled those who attended. The Electrical Building, filled with the newest devices that ran on battery and centrally provided current, was packed with fully functioning technological marvels. Could there be any doubt as to which way the world was moving? Yet electrical progress continued to advance uneasily.

At least part of the problem arose from the chaos surrounding electrical engineering itself. Even by the 1880s, when electrical devices were multiplying at a rapid rate, there were few standard solutions to similar engineering problems in different industries, such as telegraphy and electroplating, and little formalization of the kind that existed in other fields, such as civil or mechanical engineering.

Edison realized this even as his own electrical devices were coming to market. Meeting with Columbia University officials in the 1880s, he discussed the idea of an electrical engineering program. Edison was even willing to donate some equipment, including a dynamo he had used at a Paris exhibition a few years prior. The university officials seemed intrigued, even welcoming of the idea, but only if the renowned Edison was willing to establish the program with his own money. At that juncture the idea was dropped.

Years after electricity was accepted as safe, it was still seen as a luxury as well as a hallmark of progress. This is keenly apparent in the works of writers whose lives spanned the nineteenth and twentieth centuries. In F. Scott Fitzgerald's *The Great Gatsby,* the narrator, Nick Carraway, continually comments, nearly compulsively, on the lavish lighting of Gatsby's home. Gatsby stares at the green electric light that marks his love interest's home across the water; later, after Gatsby's death, Carraway reads an ancient self-help schedule Gatsby had written for himself as a child, "Study electricity, etc. . . ." And in Eugene O'Neill's play *Long Day's Journey into Night,* the family patriarch, James Tyrone, repeatedly lectures his sons about the cost of electricity, "I told you to turn out that light! We're not giving a ball. There's no reason to have the house ablaze with electricity at this time of night, burning up money!" In *Dynamo,* O'Neill sets his characters struggling between religious belief and technology, specifically electricity, actually naming one of his characters Light.

. . .

LIKE THE ELECTRIC LIGHT, THE telephone was one of those devices that seemed destined for invention. What is unique is the vast number of engineers working simultaneously toward the same end. Chief among these was Elisha Gray. Based in Chicago, Gray was already considered a serious inventor, holding patents for technical devices that enhanced telegraph systems, such as improved relays. At the time, this was no small accomplishment. Telegraphy was big business and even small enhancements were worth a fortune. Gray had already cofounded Graybar—named after himself and his partner, Enos Barton—and by the early 1870s, formed Western Electric.

For Gray, the telephone was another form of telegraph, one that would transmit sounds rather than simple clicks. In the early 1870s, he developed a telegraph that could actually transmit different sounds, each played by a separate telegraph key. By all accounts, it was a unique instrument, but of very little use in the telegraph business. It was shortly thereafter that Gray came up with the idea of a liquid microphone and primitive speaker.

Unfortunately for Gray, his lawyer submitted a patent caveat application on the same day—February 14, 1876—as Alexander Graham Bell's lawyer submitted his patent for the telephone. Gray's caveat, which is like a patent place holder, would have given Gray the credit. However, Bell's patent was approved first, and what followed was a convoluted two-year court case with accusations of bureaucratic corruption that included everything from putting Bell's paperwork quite literally on the top of the stack to allowing the young inventor to sneak a peek at Gray's documents prior to filing.

Just whose patent papers arrived first remains in dispute, though Bell finally won the case along with his place in the history books. As for Gray, he continued inventing, eventually coming up with the "telautograph," an ingenious device that employed small motors to send written messages along telegraph lines.

Even more intriguing than the Bell and Gray case was that of

Antonio Meucci. An Italian inventor, engineer, and political activist, his wanderings eventually landed him in Staten Island. Caring for his invalid wife, he began experimenting with electrotherapies to cure her arthritis, then turned his attention to ways of communicating with her from different parts of the house. Long before either Bell or Gray had filed their patent applications, he developed a crude working telephone that some date back to the 1840s or 1850s. Living in poverty and desperately trying to find financial backers for his invention, he was unable to pay even the small fee for a caveat.

Like Gray, Meucci took Bell to court. The case dragged on for nine long years before quietly fading away.

Except for a few particulars, such as an influential father-in-law, Bell would have seemed the long shot in the race to develop the telephone. Although passionate about science from an early age and something of an amateur inventor—the Scottish-born Bell earned a living not in science or industry, but teaching deaf pupils in Boston. It was there that he met his wife, who studied under him as a deaf student.

Following the path set out by other inventors, Bell's first design for the telephone resembled a musical instrument rather than the device we know today. The idea was to create something akin to a music box that could vaguely duplicate speech. Looking at the telephone as part of the telegraph's natural technological evolution, it made perfect sense to continue along the path of a mechanical system that performed some form of electromagnetic work, such as activating a telegraph key.

According to legend, when his assistant and skilled machinist Thomas Watson plucked at one of the reeds attached to a spring they were testing, the sound that came over the wire was far more detailed than anticipated. What Bell heard was the interruption of the electromagnetic field along the wire disrupted by the plucked reed, then reproduced at his end. By chance, the line had been active with a constant flow of current because either Bell or Watson had turned a screw too tightly.

It took Bell months to refine the system, and he went so far as to file a patent prior to building a working model. Eventually, he got his "speaking telegraph to work," but just barely. The first phone was an odd-looking thing, later nicknamed the "Gallows telephone" for its appearance. The way it worked was simple—someone spoke (or yelled) into a megaphone-shaped microphone, which caused a small membrane at the bottom to vibrate with sound waves. The membrane was attached to a thin rod in a metallic cup of acid with one battery-powered wire attached. Each time someone bellowed into the megaphone-like speaker, the resistance on the line changed with the vibrations of the voice as the rod moved up and down. An identical unit at the other end of the line reversed the process, essentially decoding the electrical impulses back into sound vibrations.

The Gallows telephone really had no practical application except that it showed proof of concept. However, compared to the simple opening and closing of a circuit to activate an electromagnet—which is how a telegraph functioned—Bell's device was revolutionary. What he succeeded in doing was translating fairly complex sound waves into something mechanical and then electrical in a way in which they could be retranslated back into sound waves. It was, by all standards of the day, a very neat trick, but it wasn't until the metallic cup and its acidic mixture were replaced by a magnet and a soft iron bar in the center that vibrated via a membrane to change resistance on the line that the telephone became a practical device.

In his patent as well as in correspondence, Bell referred to his invention as an "Improvement in Telegraphy." What he had done, of course, was remove cutting-edge communications technology from the realm of the professional and put it in the hands of the consumer. There was no longer a need for a third party—the telegraph operator—for two individuals to communicate over long distances, a fact not lost on Western Union.

. . .

WITHIN TWO YEARS OF HIS patent filing, there were 10,000 Bell phones in the United States, a number that would grow to more than a quarter million by the early 1890s after Bell's original patents expired. Perhaps even more impressive than the growth of the phone was the sheer amount of litigation it generated. Over the years Bell was forced to defend his patent against hundreds of lawsuits—600 by some counts—winning every one.

Telephones of all varieties began to come on the market. In some notable cases, households stored batteries, either a dry cell or a Leclanché cell, in the telephone box itself. Telephone employees would visit the homes of subscribers to service the batteries, topping off the electrolyte or changing it, as needed. It was an unwieldy system, to say the least, and with the introduction of line current provided from a central station, it finally ended.

Despite its explosive growth, there was still some puzzlement over the telephone's place in society. Very early telephone promoters seemed to genuinely believe their revolutionary device had a future almost exclusively in business communications. For them the telephone was a serious piece of technology obviously intended for serious purposes. Users were actually discouraged from tying up the limited, though expanding, resources of telephone technology with trivial chitchat. Women, in particular, were seen as the worst potential offenders of such technological abuses.

It was only in the first few years of the twentieth century that phone companies began aggressively courting consumers with promises of convenience. The phone, according to Bell executives during this time period, was a wholesome, family-friendly piece of technology, capable of making a homemaker a more efficient domestic manager. Myths around the phone also began to spring up. In one case, editors of a paper warned phone owners not to communicate with the sick for fear of contracting disease over the phone lines.

Once it became apparent that consumers actually wanted telephones, the companies wasted little time in marketing them through advertisements and stories planted in papers and magazines with the help of friendly editors as well as public demonstrations. And, too, the industry began to come up with new uses for the phone, offering what amounted to broadcasts of news, weather reports, and even concerts.

OF COURSE, THERE WERE SOME who didn't need to be convinced of electricity's potential, even if they didn't fully grasp the concepts involved. And it was these folks who attracted the confidence men. Medical quackery continued unabated, seemingly paralleling every legitimate scientific and technological advance. With the availability of relatively inexpensive power sources and the public's newly acquired blind faith in science, bad medicine took off in a big way. Why shouldn't the mysterious power of electricity—which was able to send messages thousands of miles in the blink of an eye or transmit actual voices—cure the simple ailments of the body, such as poor eyesight, depression, sexual dysfunction, gout, and irregular bowels?

In 1871, a New York City doctor named Albert Steele announced that his experiments concluded beyond all reasonable doubt "... that man is but an electrical machine and that disease is simply a disturbance or diminution of electrical forces in the system."

Needless to say, the good doctor never elaborated on just what those experiments might have been, though they were no doubt highly scientific and far beyond the understanding of those not specially schooled. Never mind the boring details. Anything, no matter how far-fetched, seemed plausible when it came to electricity and science.

For Walt Whitman, who wrote the rhapsodic poem "I Sing the Body Electric," the mysterious force of electrical current was a poetic metaphor, though for men like Steele and his loyal followers, the body really was something very close to an electrical power grid. With these claims came an implied understanding of the body's mysterious electrical circuitry. Anyone reading Steele's material couldn't help but assume that the good doctor possessed intimate knowledge of the body's wiring schematic.

And Steele wasn't alone; quackery dressed up as science found an eager population of believers as charlatans of all stripes quickly jumped on the technological bandwagon. For some of these frauds the human body was portrayed as a giant "galvanic cell," an idea that weirdly harkened back to Galvani's debunked theory of animal electricity, then reemerged awash in computer-generated special effects in science fiction form in the hit movie *The Matrix*. One popular nineteenth-century lecturer, a J. H. Bagg, related the story of a woman who was mysteriously electrically charged by the northern lights and was able to shoot sparks from her fingertips, at least for a little while. Apparently she was not only a battery, but a storage battery.

An entire industry sprang up catering to electromedicine. Medical supply houses began turning out battery-powered equipment for electrotherapies for doctors. With prices starting at $10.00 and progressing upward to $25.00 or more, they certainly looked like serious pieces of medical equipment, mounted in handsome polished wooden cases with shiny brass fittings and packed with complex wiring. They were also built for portability, for house calls. One of the best-known manufacturers, Jerome Kidder, combined batteries with small induction coils to boost the charge, while other versions included small hand-crank generators.

At a time before the FDA, it didn't take long before the same type of electrotherapy spread to home use. These devices ranged from the laughable to the absolutely frightening. Some of the machines simply generated a slight charge as patients gripped two conductive handles to fill them with the miraculous electrical healing power. However, one product, marketed by Professor W. R. Wells, included everything an electrotherapy novice needed to get started in the pri-

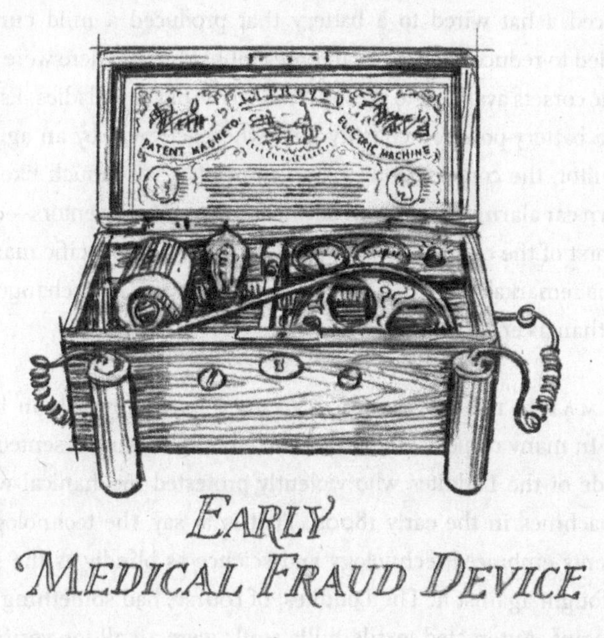


vacy of their own home, including detailed instructions on how to mix the chemicals for the battery. And for those really serious about their electrotherapy treatments, the good Professor Wells offered an optional set of medical instruments and probes designed to apply medicinal current to the eye and throat, as well as the vagina and rectum, with pinpoint accuracy. Later, wily inventors, ever watchful for new markets, began selling vibrating probes to "restore vigor," eventually giving rise to an entirely different battery-powered industry outside the medical profession.

The list of dubious electrical devices marketed is nearly endless. In England, one inventor came up with a cake of soap that was advertised as providing an electrical charge when bathing. Available details are sketchy, but it seems that the manufacturer claimed the soap released an acid that interacted with metallic electrodes—probably zinc and copper—that turned the bathtub into a battery. A few years later, as batteries began to shrink in size, a milliner produced a hat wired to a battery that produced a mild current intended to reduce headaches and prevent baldness. There were also electric corsets available to preserve the virtue of young ladies. Essentially a battery-powered chastity belt, when activated by an aggressive suitor, the corset let out a loud siren blast very much like our modern car alarms. As with all new technology, the inventors—even the worst of the charlatans—targeted concerns of a specific market. What is remarkable is just how little those concerns have changed in more than a century.

THE MARKET FOR THESE FRAUDS resided mostly in large cities. In many respects, those gullible city dwellers represented the flip side of the Luddites who violently protested mechanical weaving machines in the early 1800s. That is to say, the technological adherents embraced technology and science as blindly as the Luddites fought against it. The Luddites, of course, had something of a valid point. Automated textile mills really were an all too verifiable cause of unemployment and poverty in some regions of England. Conversely, the impassioned technophiles typically possessed little real understanding of electrical technology or its limits.

Even the most unlikely claims required only the slightest patina of science to attract followers among a growing middle class eager for a place in the vanguard of progress. Very often, all that was needed to capture the loyalty of paying customers was a serious-looking piece of equipment and a hoaxer masquerading reasonably well as a learned man of science.

Perhaps one of the most enduring of these devices was the "electric belt." First introduced in the 1870s by the British inventor J. L. Pulvermacher, but later copied by dozens of manufacturers, the belts sold briskly well into the twentieth century, promising to increase vigor, improve circulation, and enhance excretion. Wholly ineffective, but technologically ingenious in their own way, the first generation of electric belts featured multiple wooden rods coiled with zinc and copper wire to form a crude battery. Users were instructed to soak the belt in diluted vinegar and then wear it under their clothing against the skin. Of course, it must have been working, since there was a very noticeable tingle of electric current. One style of belt even included a cup to hold the man's scrotum in coils of vinegar-soaked zinc and copper while another popular model consisted of a chain arrangement to be worn around the chest.

The belt, as well as the good Professor Pulvermacher's variation of chains, gained a wide following. Not surprisingly, it even worked its way into nineteenth-century literature. In *Madame Bovary*, when Homais, the hopelessly pretentious pharmacist, mourns Madame Bovary's death, he throws himself into the fashionable chic, feeling the dead woman's influence from beyond the grave. "He was enthusiastic about the hydro-electric Pulvermacher chains; he wore one himself, and when at night he took off his flannel vest, Madame Homais stood quite dazzled before the golden spiral beneath which he was hidden, and felt her ardor redouble for this man more bandaged than a Scythian, and splendid as one of the Magi."

There were debunkers of these frauds at the time, but they faced an uphill battle. In 1892 when the *Electrical Review* sought to expose the charlatans of electric belts, England's Medical Battery Company, which made the popular Harness Belt, took the publication to court for libel. The suit was unsuccessful.

One of the more ingenious, if not dubious, uses of battery–powered electrotherapy was employed to torment at least one member of the press. Jacob Riis, the journalist and photographer who stirred America's conscience to the plight of the poor in his book *How the Other Half Lives* (1890), detailed his own experiences with electrotherapy in a later compilation of his work called *The Making of an American* (1901).

> I remember well when the temptation came to me once after a quiet hour with Police Commissioner Matthews, who had been telling

me the inside history of an affair which just then was setting the whole town by the ears. I told him that I thought I should have to print it; it was too good to keep. No, it wouldn't do, he said. I knew well enough he was right, but I insisted; the chance was too good a one to miss. Mr. Matthews shook his head. He was an invalid, and was taking his daily treatment with an electric battery while we talked and smoked. He warned me laughingly against the consequences of what I proposed to do, and changed the subject.

"Ever try these?" he said, giving me the handles. I took them, unsuspecting, and felt the current tingle in my finger-tips. The next instant it gripped me like a vice. I squirmed with pain.

"Stop!" I yelled, and tried to throw the things away; but my hands crooked themselves about them like a bird's claws and held them fast. They would not let go. I looked at the Commissioner. He was studying the battery leisurely, and slowly pulling out the plug that increased the current.

"For mercy's sake, stop!" I called to him. He looked up inquiringly.

"About that interview, now," he drawled. "Do you think you ought to print—"

"Wow, wow! Let go, I tell you!" It hurt dreadfully. He pulled the thing out another peg.

"You know it wouldn't do, really. Now, if—" He made as if to still further increase the current. I surrendered.

"Let up," I begged, "and I will not say a word. Only let up."

He set me free. He never spoke of it once in all the years I knew him, but now and again he would offer me, with a dry smile, the use of his battery as "very good for the health." I always declined with thanks.

Not only did the belts and electrotherapy gain in popularity, they also drew in some unlikely promoters. Perhaps the strangest and most unlikely was Henry Gaylord Wilshire. Born in Ohio to a prominent family, he dropped out of Harvard and moved to Southern California in the 1880s, eventually making his fortune in real estate—Wilshire Boulevard in Los Angeles is named for him. If nothing else, he was a man of widely varied and very often conflicting passions. A ruthless real estate tycoon who ran for Congress as a dyed-in-the-wool socialist, Wilshire attracted a high-profile salon of radical intellectuals and writers that included H. G. Wells, George Bernard Shaw, and Upton Sinclair.

Then in 1925 he began promoting the I-ON-A-CO electric collar, an electromagnetic device that very much resembled a horse harness. The collar was based on the extraordinarily dubious theory that an electromagnetic field somehow interacted with the body's natural iron content to restore health. According to most accounts, Wilshire was genuinely sincere in his belief that the belt provided medicinal benefits and even enlisted his friend Upton Sinclair to promote the thing. Wilshire himself not only invested heavily in the thing, but took to the road carrying with him all of the credibility of a millionaire. By the time the I-ON-A-CO craze petered out in the late 1920s, thousands of collars had been sold and tens of thousands of people treated in storefront clinics.

Even those who should have known better apparently fell for the miracle cures promised by electrotherapy. In February of 1887, *Electrical Review* published a story headlined "An Electric Treat," reporting that congressmen were sneaking off to the basement of the capitol building to fill themselves with an invigorating dose of electricity from a device rigged up in the boiler room. And, a few years later, Sir Thomas Barlow, president of the Royal College of Physicians in London, was advocating "electric cocktails" that consisted of a moist sponge fitted atop a battery and swabbed across the face.

"When you meet a friend don't offer him alcoholic stimulation; treat him to an electric cocktail," Barlow was quoted as saying. "You do not get, after electric stimulation, the injurious reaction that always follows a dose of alcohol." Something may have been lost in translation, because while the electric cocktail actually did come into fashion, it was prepared not with a battery and sponge, but with standard alcohol and a "trifle of sugar." A probe fitted with a platinum element connected to a battery heated the mixture. "It promises to be a fashionable winter beverage, and can be made cold or hot," the *Electrical Review* reported in 1885.

# 9

## *Genius by Design*

> *"Electricity: carver of light and power, devourer of time and space; bearer of human speech over land and sea; greatest servant of man—yet itself unknown."*
>
> —*Charles W. Eliot, inscription on Union Station, Washington, D.C.*

The electrical infrastructure proved painfully slow in its expansion, particularly when compared to the telegraph. As late as 1917, only about 24 percent of the homes in the United States had electricity. And, even more surprising, by 1925, decades after Edison began lighting New York from his Pearl Street power plant in the early 1880s, few rural residents had electrical power—though cities were nearly fully electrified.

In a world not yet wired for electricity, which was still seen as something of an urban novelty rather than everyday convenience, Edison led the way in developing new battery-powered products. Among the first and perhaps oddest of these was the electric pen. Developed in Edison's shop in 1875, it was among the first consumer products that featured an electric motor. The pen was a somewhat standard-sized shaft with a small electric motor mounted on top. When switched on, a tiny reciprocating needle perforated the paper with thousands of tiny holes as the user wrote, creating a stencil. The paper was then fitted into a small desktop press and copied by running ink over it with a roller, like an old-style printing press or silk screen. The ink would flow through the holes made by the needle to produce a copy.

Edison had high hopes for the pen, promoting it as a labor-saving device for use in offices. He was not completely wrong—copying documents was a time-consuming task in the 1800s. Marketed as "Edison's Autograph Press and Electric Pen," the device sold for $30.00 (about $600.00 in constant dollars). It might have achieved some level of immediate success if it had not been for the battery. Businessmen, it seemed, did not want to deal with the mess involved in lead acid batteries. Businesses, which were conservative by nature, had been resisting the typewriter for years. By some counts, there were more than fifty different versions of the typewriter created and patented over the years, all of them finding the most limited use. Around the same time as Edison introduced his pen, E. Remington and Sons, the manufacturer

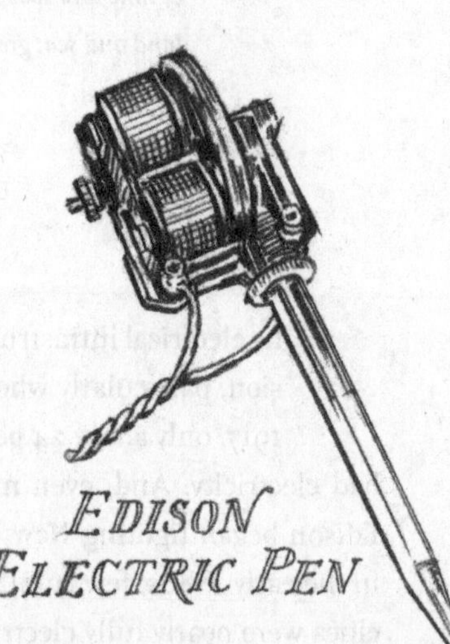

of firearms, began marketing a typewriter that finally met with some success. What chance did Edison's pen, with its mysterious noisy motor and messy batteries, have in business offices?

Edison's solution was a total redesign of the batteries that powered the pen. Still using "wet cells," he reengineered and sealed the casings so they were not easily spilled, then redesigned the entire unit, allowing owners to take it apart easily for cleaning. Finally, he added a small lever that lifted the electrode to keep the battery from running down when not in use.

Even with these improvements, the pen was not the resounding success Edison had hoped. However, it did find a few fans, most notably Charles Dodgson, better known as Lewis Carroll, the author of *Alice in Wonderland*. A few years later, the same stenciling principle as the electric pen was applied to the typewriter, which was finally catching on. The A. B. Dick Novelty Company developed a mimeograph system with Edison's help that utilized a stenciling paper that could run off copies on a specialized printer. Using the best-known technological brand name in the business, the firm eventually marketed the product as the Edison mimeograph.

The pen did not vanish entirely. In the 1890s, nearly two decades after its introduction, a New York City tattoo artist, Samuel O'Reilly, modified the electric pen to create the first modern tattoo machine, significantly shortening what had been a long and much more painful process.

BATTERIES WOULD CONTINUE TO BEDEVIL Edison right up through the turn of the century and beyond. Even as his phonograph was taking off as a popular consumer product, customers balked when he replaced the standard mechanical models that required spring winding with a new battery-powered system. Early experiments had shown that spring motors, similar to those used in clocks or music boxes, could not produce the same kind of steady rotation needed to provide clear audio from the wax cylinder. Electric motors powered by batteries were the best engineering solution, but the public was stubbornly resistant.

Perhaps it was because consumers had low expectations for sound quality from the outset. And, too, the new battery-powered technology not only required regular maintenance to change the electrolyte, but it was significantly more expensive. Just as the home entertainment market was progressing beyond the upright piano in the parlor, consumers were probably price sensitive when it came to such an obvious luxury item. At a time when the battery-powered phonograph was selling for $100, cheap spring-driven machines sold by competitors flew out of stores for $25.00 or less.

By the turn of the century, Edison somewhat reluctantly designed a spring-powered machine that sold for $10.00, and sales took off. In 1898 he sold some 14,000 inexpensive spring-powered machines, but only a few more than 400 battery-powered units. The general public was not ready for batteries as a power source.

And most batteries were not ready for the general public. With relatively few consumer products to actually power, battery technology stalled and advanced uneasily, unable to get away from the wet cell. Then, two French chemists, Felix Lalande and George Chaperon, developed what would become known as the Lalande battery. The small unit was housed in a standard earthenware container and used zinc with compressed copper oxide along with a caustic electrolyte that was alkaline, rather than acid, based. Scientists had discovered that alkalines were not subject to polarization like batteries with acidic electrolytes and

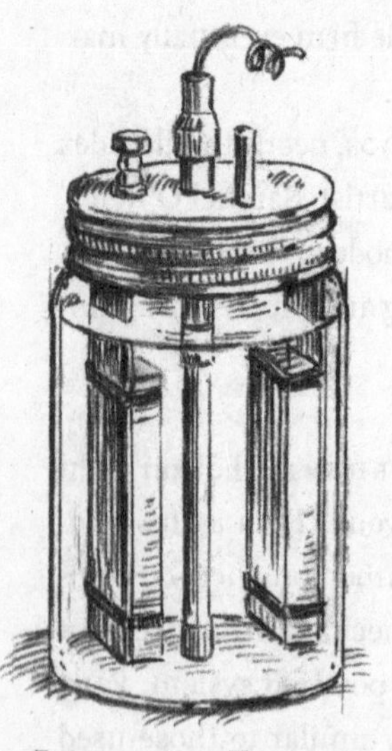

LALANDE BATTERY

still provided a steady current while sacrificing only minimal power. The Lalande battery not only produced a current strong enough to power simple motors, but also required little maintenance outside of changing the electrolyte.

The first version of the Lalande was a basic telegraphic cell, porous pot and all, intended for use by professionals. Although it was somewhat familiar-looking because of its container, the telegraph industry turned up its collective nose at the new battery. Polarization wasn't a problem for professional battery men accustomed to dealing with it through long-established routines. The professionals' problem with the Lalande battery was that it produced slightly less juice than acid-based batteries already in use.

Not discouraged, Lalande redesigned the package, substituting a sealed porcelain container for the standard porous pot. The new design didn't do anything to boost the power output, but it did produce a battery that could be used in the home. For a brief time it even enjoyed a limited run powering electric lights.

Despite his unwavering belief that central stations were the way to go when it came to electric lighting, Edison also maintained high hopes for the battery as a power source for consumer products. The newly designed Lalande battery seemed a reasonable solution for those potential customers without electricity in their home, that is to say, the vast majority of the population. They were easy to use and maintain, and eventually Edison marketed the Edison-Lalande for a variety of devices, including phonographs. A little under eight inches high and four inches in diameter, the batteries were not only durable, but also powerful enough to put out in the field unattended for extended periods, making them ideal for applications such as powering train signals.

IN 1899 EDISON UNDERTOOK WHAT would turn out to be his last great project. As automobiles began to catch on, he sought to make a battery-powered car. Automobiles had captured the public's imagination. As Edison correctly judged, they were about to become the next big thing. This would require a long-lasting, lightweight battery capable of recharging.

At the time, the dominance of the gasoline-powered car was far from a certainty, and Edison truly believed the internal combustion engine was nothing more than a bridge technology that would eventually lead to an electric car. In fact, he may have been right, but he badly misjudged just how long that bridge would endure and the difficulty in creating a durable, lightweight battery.

Electric cars were nothing new. They had been in use for years. The problem was in the weight of the batteries required to power the car. The Electrobat, for instance, introduced in 1894, weighed more than 4,000 pounds, a whopping 38 percent of which was made up of batteries—though it could go between 50 and 100 miles on a single charge.

To create a lightweight, sturdy, and easily charged power source would take Edison more than a decade of research and thousands of tests. In keeping with his proven methodology, he dispatched teams of assistants to read and paraphrase virtually everything published on primary batteries, both in the United States and Europe, including patents and technical journals going back decades. Still, Edison's own development process took thousands of experiments with hundreds of different compounds until he finally hit on a viable alkaline battery.

When chided for the many battery experiments in which he failed, he is reputed to have answered, "No, I didn't fail. I discovered 24,999 ways that the storage battery does not work."

At one point, Edison, who was never shy about promoting his products prematurely, announced in an interview,

> These batteries will run for 100 miles or more without recharging. They can be charged in a few hours. They require no attention for all that is needed to replenish the liquor [*sic*] is to pour in a little water now and then to take the place of that which has evaporated. I do not know how long it would take to wear out one of the batteries, for we have not yet been able to exhaust the possibilities of

one of them. But I feel sure one will last longer than four or five automobiles.

That particular battery failed initial testing. Another battery passed the preliminary tests and was soon rushed into production.

Following a blueprint of his past successes, Edison arranged for the batteries to be used in vehicles for a number of high-profile companies, including Montgomery Ward, the Central Brewing Co., and Tiffany & Company. Unfortunately, the battery had a flaw that sent Edison and his team back to the drawing board again.

In the end, the alkaline storage battery Edison finally perfected had virtually no chance of gaining popularity among consumers no matter how reliable it may have been. Ford's Model T, introduced in 1909 along with its reliable internal combustion engine, had become the standard for consumer autos. Henry Ford, who had once worked for the Edison Illuminating Company's generating stations, had beat Edison at his own game. He had even come up with the pithy quote, "The customer can have any color he wants so long as it's black," he was reported to have said about the Model T.

Not about to let those years of research go to waste, Edison set about finding new uses for his alkaline battery, designing a wide array of devices it could power, from railroad signals and switches to ship lighting and miner's lights. Eventually, it became one of the most profitable divisions of Edison's empire.

However, the long years spent in battery development may have also distracted the Wizard of Menlo Park from other inventions coming on line at the time. He rejected radio, calling it a "craze" and took special pains to explain that " . . . there are several laws of nature which cannot be overcome when attempts are made to make the radio a musical instrument." For years he resisted building a phonograph with a radio integrated into the unit, seeing the two technologies in competition for consumers' attention, even as his distributors and customers demanded just such a product.

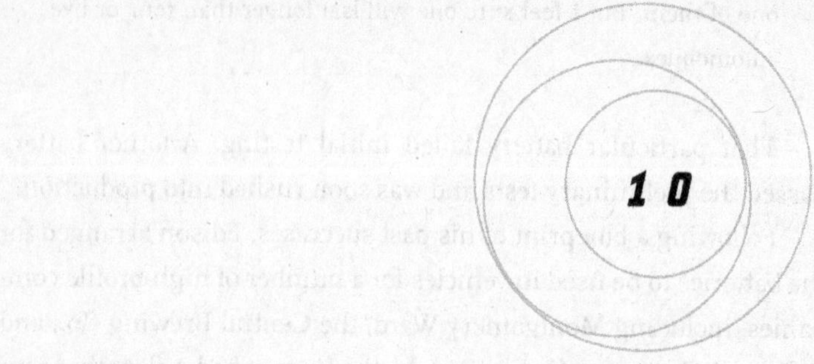


# Victorian Age of Discovery

*"To the electron—may it never be of any use to anybody!"*

*—Joseph John Thomson*

The late nineteenth century was an exciting time for science. In the 1880s, the German physicist Heinrich Hertz proved the existence of radio waves—called Hertzian waves—emanating from simple electrical sparks generated from a battery. Wilhelm Röntgen (sometimes Roentgen) discovered radiation in fluorescing glass tubes. And in England, Joseph John Thomson's work at Cambridge led him to study the effects of electricity and magnetism on gas that would unlock the secrets of the atom.

Like Faraday, Thomson (J. J. to his friends and colleagues), who secured his reputation for genius at an early age, came to science by unlikely chance. His father, a bookseller outside Manchester, had pushed for his son to go into engineering. But, unable to pay the required apprentice fees, Thomson was enrolled at nearby Owens College (now the University of Manchester) to study math. Math was better than nothing and perhaps some good would come of it. As it turned out, young Thomson's brilliance at math was quickly recognized and within a short time he was shipped off on scholarship to Cambridge, where his genius came into full bloom.

Elected Cavendish Professor of Experimental Physics before the age of thirty, Thomson was a quiet, unassuming sort, gentle and bespectacled; he was more comfortable with mathematical proofs scrawled on a blackboard than with experimentation. In many ways, he was the mirror opposite of Faraday, who honed his fine motor skills at the bookbinding trade of his youth and became the master experimenter, but struggled with higher mathematics. Conversely, Thomson breezed through the equations of complex math, but was extraordinarily, nearly comically clumsy around glass tubes and beakers. According to legend, his students stood fearfully by whenever he approached their experiments, afraid that he'd topple their precisely arranged work. Paradoxically it was a fairly complex piece of lab equipment that was essential to Thomson's breakthrough experiment.

It had been known for some time that electricity boosted to high voltages with an induction (or Ruhmkorff) coil behaved oddly when exposed to gas in a sealed glass container. The glass glowed luminously. Scientists had been studying the phenomenon as far back as 1858 with little to show for it other than beautifully lit cylinders of glass. What exactly was happening inside the tubes remained a mystery. The glow originated in the negatively charged cathode and traveled the short distance to the positive anode, but beyond that point very little was known. Some physicists theorized that perhaps the glow was caused by an unidentified interaction with electricity that produced light waves. Thomson conceived of an experiment to settle the matter. When he sealed a glass tube with two metal plates on each end and connected the plates to a battery and induction coil, the tube glowed—projecting the mysterious light from the negative to the positively charged plate. What Thomson and a few others had built was a cathode ray tube—essentially a very primitive version of the picture tubes once widely used in televisions and computer monitors. Thomson theorized that the mysterious glow was not caused by light waves, but rather by negatively charged particles pouring off the flat cathode (negatively charged metal) at the end of the tube.

They were, he guessed, attracted to the positively charged anode. If a magnet were placed nearby, the electromagnetic field would bend the flow of particles—very much in the same way that kids distort the picture on a television's image by placing a magnet near the picture tube. The positive pull of the magnet attracted the negatively charged particles.

Others had tried the same experiment but failed, primarily because their apparatus had not sufficiently cleared the tube of all the gases. Thomson's experiment succeeded. Not only did it succeed, but he could measure the ratio of charge to mass of the particles by the way the electromagnetic field bent the beam of light.

The results, by any acceptable explanation, were a little wacky. All the data indicated that whatever was being projected was much smaller than a hydrogen atom, the smallest, lightest matter known. Thomson had discovered the electron, the first known subatomic particle, which he dubbed the "corpuscle." Other physicists eventually proved him right with their calculations. It was only later that the name changed to the less visceral "electron"—a name Thomson stubbornly rejected until sometime in 1914—eight years after winning the Nobel Prize for its discovery.

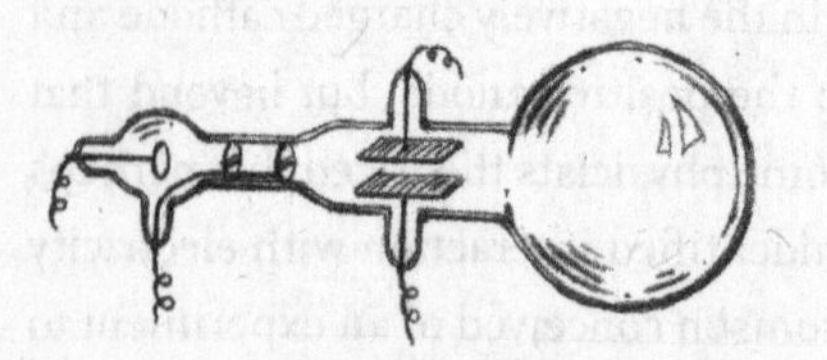

EARLY CATHODE RAY TUBE

Edison had witnessed a very similar phenomenon years earlier. When experimenting with his lightbulb, he inserted an extra electrode into the bulb and noticed the interior blackening with the discharge of electrons. The phenomenon, which he called "the Edison effect," was largely ignored by the inventor. It wasn't until late in his life when he was actively seeking acceptance by the scientific community that he would tout his discovery. "I was working on so many things at the time, that I had no time to do anything more about it," he said.

"To the electron—may it never be of any use to anybody!" became Thomson's favorite toast.

AT AROUND THE TIME LALANDE and Chaperon were refining the design for their battery in France, the German chemist Carl Gassner patented what came to be known as the "dry cell." In a simple variation on the Leclanché battery, Gassner mixed ammonium chloride with plaster of paris and some zinc chloride, and then sealed it in a zinc container.

It was an ingenious design; the zinc can that housed the battery also served as the negative electrode. Pumping out a steady 1.5 volts, the advantages of Gassner's design over standard "wet cells" were immense, particularly for the consumer market. It didn't spill or require maintenance, could be mounted in virtually any position, and was more or less every bit as reliable as anything on the market, even the Lalande-Chaperon design. And, too, because it was a solid, Gassner's battery could easily be scaled down to virtually any size. Gassner wasted no time in patenting his new battery throughout Europe and the United States in 1887. In the 1890s, the Cleveland-based National Carbon Company (later known as Eveready, and then Energizer), which had been manufacturing Leclanché wet cells, made some basic modifications to Gassner's original design and began marketing the batteries under the brand name Columbia dry cell. An instant success, the Columbia dry cell measured some six inches long and, like the original, produced a steady 1.5 volts.

Cheap to manufacture and easy to use, the sealed, carbon zinc battery with an acidic electrolyte was the first mass-produced consumer battery in the United States. Finally, here was a compact, durable battery with which to power all manner of devices.

Thanks to the new battery, electric power was now truly portable and ready for the consumer market. A whole slew of novelty companies began selling strange and often gaudy devices, sometimes using the Columbia dry cell, but very often producing their own handmade batteries. One could buy electric ties and stickpins lit by tiny bulbs and powered by a battery concealed in the pocket of a suit coat. The McConnell Segar Company of Indianapolis sold the "Ever Ready Electric Walking Cane" with a small bulb mounted on the top encased in glass.

*The first mass-produced battery, the Columbia dry cell pumped out a reliable 1.5 volts to power a new generation of gadgets at the dawn of the twentieth century.*

However, even with a reliable form of energy packaged in a format acceptable to consumers, there were still relatively few practical things to power. There were electric clocks, many of them housed in extravagantly carved wooden cases, and the continued spread of doorbells. There were bicycle lights, catering to the increas-

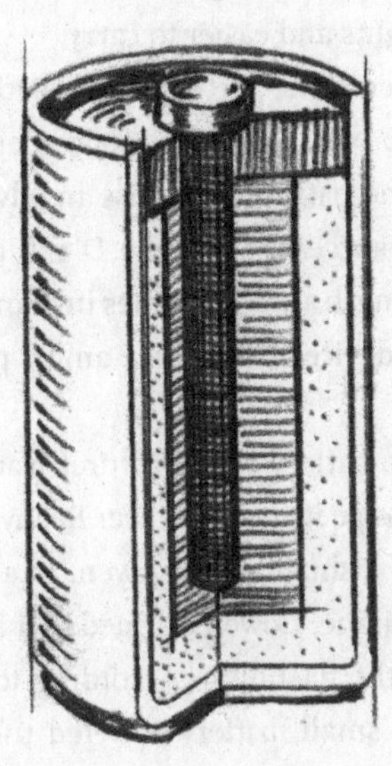


ingly popular form of transportation, and electric insoles to warm your feet. But these were mostly novelties; the average consumer could do just fine without battery-powered products.

The story of the most common of all battery-powered devices, the flashlight, remains somewhat clouded. Here is its simplest form: in the late 1800s, Conrad Hubert, a Russian immigrant, who was already marketing an assortment of electric novelties including battery-powered stickpins through his American Ever-Ready Company, teamed up with a David Missell (sometimes spelled Misell), who had worked for Birdsall Electric, which also made battery-powered novelties. Together they formed the American Electrical Novelty & Manufacturing Company to sell the flashlight along with other battery-powered products, such as bicycle lights.

In truth, there were flashlights already on the market in the form of reconfigured bicycle lights. Missell and Hubert's innovation was to house their product around what would become known as the D cell battery in a tubular design, making them lighter than the square wooden or metal cases of bicycle lights and easier to carry.

Lacking neither ambition nor nerve, the partners promoted their flashlights by giving them away to New York City policemen and then collected testimonials from the patrolmen to use in advertising. The light was an unqualified success. Since the D cell at the time measured just 2 inches in length and 1 inches in diameter, three of them could fit in the new device and provide ample power for extended periods of time.

Eventually, the firm was sold to battery manufacturer National Carbon, which would eventually change its name to Ever Ready. And it might have ended there, if not for Joshua Lionel Cowen. In a 1947 interview with *The New Yorker* magazine, Cowen claimed that it was he, not Missell, who had invented the flashlight. According to this new story, Cowen came up with a small, battery-powered tubular light to illuminate potted plants. Missell had simply removed the tube from the decorative planter to create the flashlight. Later, he explained, Missell and Hubert bought out his company for a song.

A native New Yorker and a tinkerer from a young age, Cowen

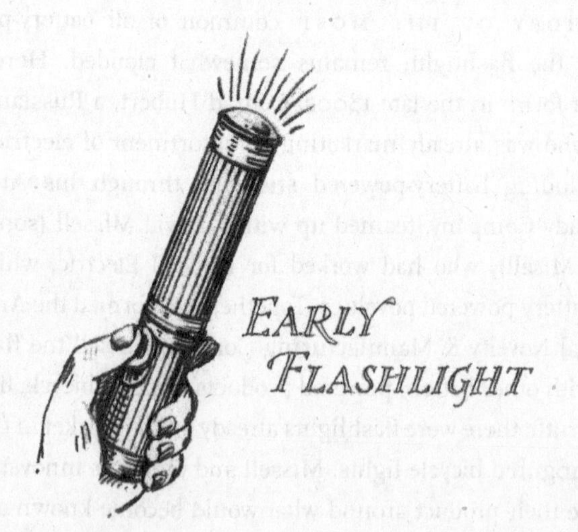

attended City College and then Columbia University for a short time before landing at a lamp factory. According to his own account, by experimenting at night he came up with a new type of flash powder for cameras at eighteen, took out a patent and found the U.S. Navy was interested in the formula for mines, eventually building enough detonators to equip some 24,000 mines.

With the contract completed, Cowen began building flower pot lights and hired Missell to sell them. Cowen recalled that he gave the idea to Hubert in 1906 after either growing bored with the flower pots or tired of complaints from customers. Hubert then, at least according to Cowen, sold half interest in the American Electrical Novelty and Manufacturing to the National Carbon Company, Ever Ready's supplier of raw materials for $200,000.

As part of the deal, Conrad Hubert remained president of the company, whose name was officially changed to the American Ever Ready Company. In addition to the company name change, the trade name "Ever Ready" became "Eveready." And by the time he died in the 1920s, Hubert had amassed a fortune said to total some $6 million.

Despite the technical flaws in Cowen's unlikely account, such as the fact that battery-powered lights could not operate for extended periods, the story might have ended with Cowen a bitter old man who gave away the invention of a lifetime. But Cowen plowed the relatively small amount of money he received from the venture back into tinkering. Within a few years, he had come up with another invention—the toy train. Using his middle name—Lionel—he went on to build a miniature railroad empire of toy trains.

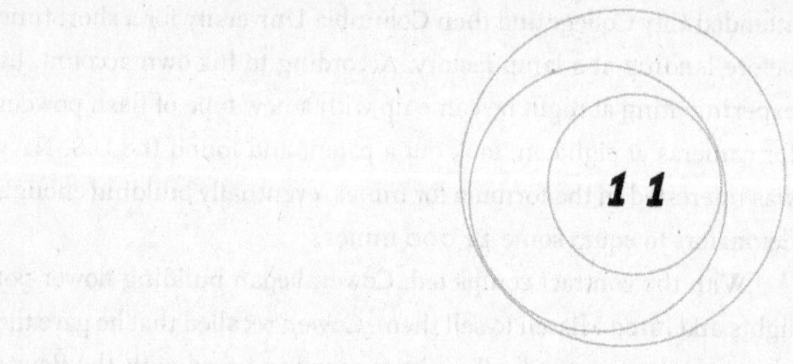

# Without Wires

*"Marconi plays the mamba, listen to the radio"*

*—Jefferson Starship, "We Built This City"*

In one version of the story, Guglielmo ("Willy") Marconi's inspiration for wireless telegraphy came from reading the obituary of Heinrich Hertz in 1894. The German physicist, using Leyden jar "batteries," demonstrated that a powerful electrical spark set off between a gap of two closely placed metallic poles emanated invisible waves that caused a similar spark in similarly gapped metal poles a few feet away. In another version of the story, he precociously read of Hertz's 1860s experiments at the age of fourteen while on vacation in the Alps and rushed back home to begin his experiments.

Hertz's sole interest seems to have been to prove Maxwell's theories correct through experimentation. One could detect the sparks received, just barely, but in Hertz's mind it was doubtful they were capable of any substantial "work." For the German scientist, just establishing their existence and giving physical form to Maxwell's elegant math was enough.

However news of the mysterious Hertzian waves reached him, Marconi was ideally, if unconventionally, suited to put them to practical use. The son of a wealthy Italian businessman and the heiress to the Jameson Irish whiskey fortune, Marconi certainly had the resources, if not the academic credentials. His education, by all accounts, was somewhat spotty, comprised mostly of private tutors along with brief stints at various schools and universities in England, Florence, and Livorno.

It was, by all accounts, a comfortable life. As the magazine *Vanity Fair* noted somewhat unkindly, later in his career, "The true inventor labors in an attic, lives chiefly upon buns, sells his watch to obtain chemicals, and finally after desperate privations succeeds in making a gigantic fortune for other people. Guglielmo Marconi invented in comfort, retained any small articles of jewelry in his possession, and never starved for more than five hours at a time." All that certainly seems to be true, at least by available accounts. Still, even from a young age, Marconi was as single-minded in his pursuit as any bun-eating inventor struggling in poverty. Close-mouthed about his work, even with his family, the young man resisted all attempts to steer him toward a more reputable career.

Blithely flunking the exam for the Naval College, Marconi continued with his experiments, much to the consternation of his father, who viewed science as somewhat less than a respectable career path. Whatever support he received at home was from his mother, Annie. Strong-willed and independent, she arrived in Italy to study opera and ended up in what was then called a "runaway marriage" to the much older, though respectably wealthy, Giuseppe Marconi. Throughout a good portion of young Willy's life, she was never far from her son, providing him with what he needed for his work and smoothing the way for his efforts. By any standard, she showed a remarkable—even uncomfortable—level of devotion. Perhaps because her own ambitions of an operatic career had been thwarted by her parents, who saw life on the stage as uncomfortably close to scandalous, she mustered the iron will to make her son's ambitions a reality.

Not surprisingly, Marconi's father proved a somewhat reluctant supporter of his son's technological inquiries, though he still provided him with the leading scientific journals and books. Devouring these texts, the young Marconi was able to gain something of an understanding of practical scientific principles. Years later in interviews, he would relate his love of Faraday's lectures, but he held a special place for Benjamin Franklin and his early electrical experiments, even going so far as to duplicate some of them on the family estate.

That a youthful Marconi should latch on to Faraday and Franklin, two master experimenters, is not surprising. His eccentric schooling did not include the advanced math required for pure science. These heroes were from an age of science without the complications of higher math. Profoundly curious about the nature of electricity and dogged in his experiments, Franklin's focus, like Marconi's, was on the potential of practical applications that could emerge from experimentation.

Setting up a makeshift lab in a dusty attic room that had once housed silkworms on his family's estate, the Villa Griffone in Pontecchio, just outside Bologna, Marconi labored within a few miles of where Galvani had conducted some of his first battery experiments at the start of the century. Experimenting on the lawns and vineyards of the estate, he discovered that Hertzian waves could travel through or over hills and trees. They could also travel distances far beyond the few yards that Hertz established to prove Maxwell's theory. Marconi quite literally brought the science out of the lab and into the wider world, though like everyone else, he had no idea how the waves traveled. Indeed, a definitive answer wouldn't present itself until the 1920s—long after regular broadcasts were more or less commonplace.

. . .

DESPITE POPULAR MYTH, MARCONI DID not emerge from his attic room carrying a new technology and present it triumphantly to a grateful world. Wireless communication had long been in the incubation stage, waiting for someone to assemble the disparate pieces through tinkering and experimentation. In this regard, Marconi was more engineer than scientist.

Years before the young Marconi even began experimenting, the eminent British scientist Sir William Crookes speculated on the possibility of wireless telegraphy.

> Rays of light will not pierce through a wall, nor, as we know only too well, through a London fog; but electrical vibrations of a yard or more in wave-length will easily pierce such media, which to them will be transparent. Here is revealed the bewildering possibility of telegraphy without wires, posts, cables, or any of our present costly appliances . . . All the requisites needed to bring it within the grasp of daily life are well within the possibilities of discovery, and are so reasonable and so clearly in the path of researches which are now being actively prosecuted in every capital of Europe, that we may any day expect to hear that they have emerged from the realms of speculation into those of sober fact.

The wireless telegraph, as it was called, like the telephone and countless other technological breakthroughs, was one of those inventions destined to come into the world sooner or later. Alexander Popov, the Russian physicist, had transmitted signals over a short distance but never published his findings, while the wildly eccentric Nikola Tesla later claimed that Marconi had made use of many of his patents, eventually launching an unsuccessful lawsuit against the young inventor. The British scientist and inventor David Edward Hughes also dabbled in the reception of Hertzian waves, which he called "ariel telegraphy" years before Maxwell published his theories or Hertz proved them through experiment. However, his work sat dormant and largely forgotten until J. J. Fahie published *A History of Wireless Telegraphy* in the early 1900s to set the record straight, in a very gentlemanly way, of course.

Even earlier, in the 1860s, an American dentist and amateur inventor in Virginia, Dr. Mahlon Loomis, apparently transmitted signals more than a dozen miles across the Blue Ridge Mountains, from the Catoctin Ridge to the Bear's Den Mountain. He even received a patent for a vaguely worded description, though his research was stalled by the Civil War, and funding tied up in Congress. Nothing ever came of his invention.

There was also Amos Dolbear, a physics professor at Tufts College, who patented a device somewhat similar to Marconi's in 1882, which led to a dispute years later over Marconi's American patent, the famous "7777 Patent." And, in India, Professor Jagadish Chandra Bose of the Presidency College succeeded in sending out Hertzian waves that rang a bell and exploded a mine, but, like Loomis, he couldn't manage to find funding for his experiments.

Even Edison had gotten into wireless transmission, devising a system that transmitted from an overhead line to a moving railroad car, though he did this through induction of two closely positioned wires rather than transmission of Hertzian waves. This was not so far-fetched an idea. For a very brief time induction was seen as having a genuine future as a form of wireless communication. In London it was discovered that telephone lines were picking up signals from nearby telegraph lines by way of induction. The esteemed John Trowbridge, who was instrumental in building Harvard University's physics laboratory and in the process brought physics out of the lecture hall and into the lab, suggested using induction communications from Europe to North America by means of "power dynamo electric machines."

Then there is the intriguing—and somewhat mysterious—case of Captain Henry Jackson of the Royal Navy, who was reported to have achieved some success with Hertzian wave transmission just prior to Marconi, though his experiments were considered state secrets. At one point, Jackson would later recall, he approached Marconi during a public demonstration and told him of the experiments, though the results were not made public at the time.

There were others scattered throughout Europe, all experimenting with Hertzian waves with varying degrees of success. What Marconi had going for him was the sheer tenacity and energy of youth along with an instinctive ability as an engineer to reconfigure what was then largely considered laboratory equipment and commercial devices into a practical transmitting and receiving system. In the same way, Edison had his teams search out pertinent patents for technology, Marconi scavenged through existing scientific instruments to find what he needed or what he could reconfigure and modify to fit his needs.

The receiver he used, for instance, was first developed in the 1890s by the French physicist Édouard Branly, a professor at the Catholic University in Paris, who had taken some metal filings and dumped them into a test tube. When hit with an electrical burst from a battery, the filings coalesced to form a kind of fragile wire capable of conducting electrical current. A few years later, the British professor Oliver Lodge discovered that the filing-filled tube reacted similarly to identify Hertzian waves in much the same way a voltmeter detects current in a line, though it could also be used as a switch or valve. Lodge called his invention a coherer. During a series of experiments, he sent signals fifty or more yards, but he never followed up on the research by seeking to extend the distance or adapt it to a commercial communications system.

It was Marconi who improved the coherer for commercial applications. Experimenting with different metallic filings, he finally hit on the combination of coarse silver and nickel powder in a vacuum-sealed thermometer tube, making it even more sensitive to electromagnetic waves, then invented a little hammer (or trembler) that worked like a doorbell to gently tap the glass and de-coalesce the filings, terminating the circuit after activation. It was an ingenious design. Each time a burst of electromagnetic energy hit the coherer, the metallic filings completed a local battery-powered circuit connected to a Leclanché-type cell that ran a standard telegraph printer. When the signal ceased, the little hammer would tap the glass and break the fragile string of metallic filings to await a new burst of waves. In many respects, the system resembled a telegraph relay that used a relatively weak signal to switch on a local circuit powered by a stronger, fresh battery.

The transmitter Marconi first used sparked across a gap of eight inches, powered by batteries supplying fifteen volts and boosted by an induction coil that sent the signal a few hundred yards. From there, he slowly worked his way to a few miles. In one interview given just a few years after Marconi debuted his system in London, the American reporter described the outfit as consisting of ninety-eight dry cells connected in parallel to eight rechargeable or storage batteries that stepped up the voltage before sending it along to a powerful induction coil that increased the voltage even more.

Later, the spark gap would grow to a whopping ten inches or more, sending the signals even greater distances with the size of the gap determining the length of the radio waves. To our modern mind, the transmission of radio waves is a silent process, though with Marconi's device they were dramatic events, the sparks noisily arcing across two posts topped by brass balls, forever giving radio operators the nickname "sparks" or "sparky."

Whitehouse may have taken some grim satisfaction in the fact that although his use of high voltage was a disaster when running current through wires, it was an absolute necessity to transmit signals through the air. A thimble-sized battery just wouldn't do the trick. To get those metallic filings to line up in a coherer, a robust burst of energy was needed. One early demonstration required more than a hundred dry cells to generate enough power for even a short-range transmission.

WITH THESE BASIC ELEMENTS ALREADY in place, Marconi had only to refine the concept to receive Morse code. However, even after he had developed a working model—proof of concept—the Italian government declared itself uninterested. This might have ended his ambitions right there, relegating him to a technological footnote and "also ran" like Bose or Loomis, but the young inventor had luck as well as support. Undaunted, he quickly turned from the Italian to the British side of the family. His mother, always his staunchest advocate, began pulling strings with her well-connected Jameson relatives in England. It was her nephew, Henry Jameson-Davis, an engineer himself, who led the charge in London for the young inventor, arranging introductions even before Marconi boarded the boat.

Marconi arrived in London only to have his apparatus broken by an overly zealous customs official, and then on June 2, 1896, filed the first patent for a wireless telegraphy device. Interestingly, it took a team of lawyers months to work out the precise phrasing for the patent application. What exactly had Marconi invented? All of the elements of the boyish Italian's device had long been in place, including the theoretical work accomplished years previously. Even the transmission and reception of Hertzian waves was not particularly unique. Scientific journals across Europe had been filled for years with experiments involving Hertzian waves. Finally, the patent was titled "Improvements in Transmitting Electrical Impulses and Signals, and in Apparatus . . ."

WITH HIS FATHER'S FORTUNE ALONG with the substantial weight of his mother's Jameson name smoothing the way, Marconi managed to arrange a series of demonstrations, first for officials at the powerful British Post Office, which controlled the country's telegraph system, then for the general public. Still in his early twenties, Marconi—who spoke flawless English—demonstrated a simple version of his device in the fall of 1896 at Toynbee Hall in the less than elite section of Whitechapel of London's East End. Home to the fictional Fagin in *Oliver Twist* and the decidedly real Jack the Ripper, Whitechapel was certainly not the Royal Institution, which was only fitting for a demonstration very much intended for the general public, rather than for scientists.

The device that Marconi demonstrated consisted of two simple wooden boxes housing the sending and receiving apparatus. When a lever was depressed on one box, a bell rang in the other. The press, which reported on the demonstration, was soon calling Marconi the "inventor of wireless telegraphy." The idea that Signor Marconi—appearing no older than a teenager—had labored away in isolation on an Italian estate to create a miraculous device was simply too good a narrative for either press or public to easily abandon. Like Edison, Marconi quickly became a celebrity inventor, giving interview after interview in the popular press.

However, not all of the press was glowing. Some feared that wireless communication could be used to set off explosions at a distance. A few reporters speculated that the young man might have developed a dangerous new weapon. And Oliver Lodge, among others, vigorously objected to Marconi's designation as "inventor of wireless telegraphy" and took steps to set the record straight. In an 1897 letter to the *Times*, he complained in the most gentlemanly manner:

> It appears that many persons suppose that the method of signaling across space by means of Hertzian waves received by a Branly tube of filings is a new discovery made by Signor Marconi. It is

> well known to physicists, and perhaps the public may be willing to share the information, that I myself showed what was essentially the same plan of signaling in 1894 . . .

The public didn't much care. Lodge had missed the point. Marconi, whom the British press habitually noted dressed as a "pleasant young English gentleman," was an instant nineteenth-century celebrity, his name indelibly linked to the invention of wireless communication. If the young Italian genius wasn't actually British, then he at least looked and sounded the part. And, too, the Jameson family was already raising money—with Jameson-Davis in the vanguard—mostly from those associated with the family's distillery business. Within a short time, the Wireless Telegraph and Signal Company (later called Marconi Telegraph) was founded with the equivalent of more than $10 million of what could be called venture capital.

In its early form, the system could send out a message a few miles at fifteen words a minute, slower than a wire telegraph. However, bit by bit Marconi was able to increase the distances, progressing from a few miles to across the English Channel. Where Marconi shone was in his development of antennas, at first grounding them (like lightning rods) and then extending them upward with the use of kites and balloons, until finally hitting on the idea of the directional antenna or aerial, which required hundreds of yards of land. It is tempting to speculate on just how much his reading of Franklin and those early colonial experiments influenced Marconi's efforts.

Each step of the way the media of the day eagerly followed his progress. Of course, the telegraph companies also kept a close eye on his advances, though with considerably less enthusiasm. Fortunes in stock stood to be lost. With their systems of cables and poles in danger of becoming obsolete, they launched something of a campaign against wireless, paying physicists and other scientists to publicly question the possibility and practicality of wireless as a means of communication.

However, the future had already arrived. By 1897, predictions for the new space telegraphy were already being made. At a lecture given at the Imperial Institute, Professor William Edward Ayrton said,

> I have told you about the past and about the present. What about the future? Well, there is no doubt the day will come, maybe when you and I are forgotten, when copper wires, gutta-percha coverings and iron sheathings will be relegated to the Museum of Antiquities. Then, when a person wants to telegraph to a friend, he knows not where, he will call in an electro-magnetic voice, which will be heard loud by him who has the electro-magnetic ear, but will be silent to everyone else. He will call, "Where are you?" and the reply will come, "I am at the bottom of the coal mine" or "Crossing the Andes" or "In the middle of the Pacific" . . .

Eventually ships would act as the perfect proving ground for wireless transmissions. This was an area not only in need of wireless communication, but one in which the established telegraph companies could not compete. Marconi's system was portable, untethered by the complex network of wire that now crisscrossed Europe and North America.

By 1898, he outfitted a yacht with an antenna and wireless set to report back on a regatta for a newspaper. It was a good publicity stunt, that is to say both the paper and Marconi benefited. This led to sets installed on the Royal Yacht and on the grounds of Queen Victoria's estate, which proved even better publicity for Marconi. The technology could have received no better endorsement than that from the aging queen. Later, Marconi would get credit for the first car phone, setting up a wireless telegraph in an enormous steam-powered car with a sixteen-and-a-half-foot antenna mounted on the roof.

In 1901, Marconi transmitted a message between the Isle of Wight and Cornwall, a distance of nearly 200 miles, demonstrating that radio waves followed the curvature of the earth. Then, in 1902, using powerful battery arrays, he transmitted his first transatlantic signal from England to Newfoundland—the letter "S" in Morse code.

. . .

JUST AS IMPORTANT AS DISTANCE, portability of wireless transmission was a fact not overlooked by the military. Naval ships, in particular, were stuck in the early 1800s, using flags for line-of-sight signaling. By 1904, during the Russo-Japanese War, wireless received its first use in combat with disastrous consequences for the Russians who were slow in adapting their strategy to a wireless world. The Russian army also bought Marconi sets to communicate with distant outposts, though one unit came close to being destroyed when a Russian Orthodox priest in Siberia insisted on blessing it with holy water.

SO FIRMLY EMBEDDED WAS THE idea of the telegraph that Marconi and others seemed to be using it as their model when casting their eye to future innovations and uses of wireless communications. One central idea for the new technology was sending a signal to a specific receiver, point-to-point communications. The concept of what would become known as "broadcasting" (borrowed from an agricultural term for casting seeds widely) sending a signal to whoever happened to possess a receiver was only vaguely considered early on, in much the same way that early IBM executives could not comfortably imagine a home use for a computer. In an interview appearing in an 1899 edition of *McClure's Magazine*, one of Marconi's engineers took a look into the future of broadcasting:

> "—any two private individuals might communicate freely without fear of being understood by others," he said. "There are possibilities

> here, granting a limitless number of distinct tunings for transmitter and receiver, that threaten our whole telephone system. I may add, our whole newspaper system . . . The news might be ticked off tapes every hour right into the houses of all subscribers who had receiving-instruments tuned to a certain transmitter at the news-distributing station. Then the subscribers would have merely to glance over their tapes to learn what was happening in the world."

However, despite fanciful predictions, what Marconi developed was still largely a device for use by professionals—excellent for communications by trained telegraphers between distant points, but with limited use for the consumer marketplace. For one thing, it lacked the essential elements of intuitive and easy operation. The subscribers Marconi's engineer imagined would have to be highly motivated indeed to operate the wireless sets and learn Morse code to decipher its messages. It would take decades before radio operation became a matter of fiddling with a few dials and switches.

NOT EVERYONE WAS WELCOMING OF the new technology. Distance was now dying at a rapid pace. Within a single lifetime the world had progressed from telegraph networks of short-distance wires to long-distance wireless transmissions. The century that began with the movement of increasingly larger objects, railroad engines and ships by way of ever-more powerful engines, now ended with the taming of the smallest of things—the electron, less than a millionth of an ounce.

"A Triumph But Still a Terror," read a *New York Times* headline in 1906. "There is something almost terrifying in the news . . . that attempts at telephoning without wires have already attained such success that scientists announce the approach of the time when man will be able to speak without any conducting wire to a friend in any part of the world," the story read, as if the physical presence of a wire provided some crucial human link.

Addressing an audience of scientists a few years before his death, Hertz said, "[Electricity] has become a mighty kingdom. We perceive [it] in a thousand places where we had no proof of its existence before . . . the domain of electricity extends over the whole of nature."

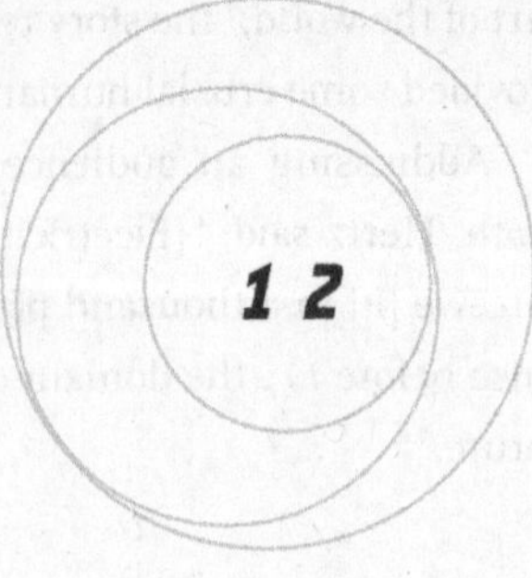


## Mass-Marketing Miracles

*"It Was Written for Boys, but Others May Read It."*
—*L. Frank Baum,* The Master Key: An Electrical Fairy Tale

On November 25, 1905, a small ad appeared in the back pages of the *Scientific American.* Placed by the Electro Importing Company, the ad was for the Telimco Wireless Telegraph Outfit. The kit was being marketed under a somewhat awkward contraction of the company's name, though the majority of readers were probably captivated by the technology itself. A hobbyist's version of Marconi's basic system, the Telimco ad promised everything an amateur needed to transmit and receive wireless signals for up to one mile, including a one-inch spark coil, telegraph key, coherer, decoherer, "catch wires," four dry cell batteries, and a speakerlike device to hear the signal. Priced at just $8.50 (a little under $200 in constant dollars), the unit was nothing short of a technological bargain. Later, department stores, such as Macy's and Gimbels also began selling the outfit, while other mail-order outfits such as Johnson Smith & Co. would market their own versions.

The ad and Telimco unit were the brainchild of Hugo "Huck" Gernsback, born Hugo Gernsbacker in Liechtenstein in 1884. Although barely twenty when he arrived in America, he had a solid educational grounding in technology along with a restless mind—certainly too restless for the staid laboratories and lecture halls of Europe and perhaps even too impatient for the laborious, often tedious work of scientific research in general.

Seeking his fortune in America with a newly designed powerful dry cell, he quickly persuaded the Packard Motor Car Company to buy the rights to his battery, then used the funds to open a store in lower Manhattan catering to a public hungry for technology. It was a brilliant concept.

America was embarking on a new century with a growing middle class that was every bit as fascinated by science and technology as the European aristocrats had been a hundred years earlier. Although Gernsback was not alone in marketing modern marvels to the masses he was certainly one of the most visible promoters, using many of the same whiz-bang gimmicks the charlatans employed to sell their medical devices and elixirs.

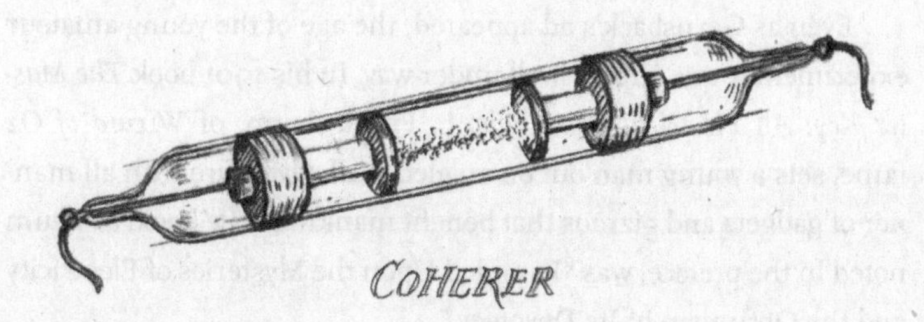

COHERER

At one point, according to popular legend, perhaps spread by Gernsback himself, The Electro Importing Company was investigated by the New York City Police Department for possible fraud. Could it be that the miracle of wireless transmission and reception were actually available for as little as $8.50? Yes, it most certainly was, Gernsback himself assured the public along with the NYPD. Marconi's miracle technology was now within reach of practically everyone everywhere.

By the turn of the century, an entire generation had grown up in the world of long-distance telegraphy and the telephone, the Wright brothers had already flown, albeit briefly, at Kitty Hawk, and the electric light, though far from a ubiquitous fixture in American homes, was no longer an illuminating novelty. Inventors, if not scientists, were celebrities in their own right with the newly minted mythologies of Edison, Morse, and others already firmly enshrined in the American psyche. Textbooks and popular magazines offered up stirring and inspirational accounts of inventors, often giving more words to the man than the machine.

Parents eager to set their children on the right path bought them chemistry sets with names like Chemcraft, produced by the Porter Chemical Company, which also sponsored Chemists Clubs. Young adult books on electricity and technology flourished, many of them written for boys and intended to instill the noble desire to invent. Who knew when the next Edison or Morse might arrive on the scene to offer up the world a truly wonderful and practical miracle?

Even as Gernsback's ad appeared, the age of the young amateur experimenter was already well under way. In his 1901 book *The Master Key: An Electrical Fairy Tale*, L. Frank Baum, of *Wizard of Oz* fame, sets a young man out on an electrical adventure with all manner of gadgets and gizmos that benefit mankind. The book, as Baum noted in the preface, was "Founded Upon the Mysteries of Electricity and the Optimism of Its Devotees."

Though now largely forgotten, *The Master Key* is a fascinating techno-literary artifact. Spanning the nineteenth and twentieth centuries, its pages are packed with early twentieth-century optimism as well as suspicion of electricity, though the doubters are quickly converted. In one early scene the young protagonist's father and mother gently quarrel about the young boy's hobby in experimenting.

"Electricity," said the old gentleman, sagely, "is destined to become the motive power of the world. The future advance of civilization will be along electrical lines. Our boy may become a great inventor and astonish the world with his wonderful creations."

"And in the meantime," said the mother, despairingly, "we shall all be electrocuted, or the house burned down by crossed wires, or we shall be blown into eternity by an explosion of chemicals!"

"Nonsense!" ejaculated the proud father. "Rob's storage batteries are not powerful enough to electrocute one or set the house on fire. Do give the boy a chance, Belinda."

THIS WAS CERTAINLY A FAR cry from the genteel instruction on natural philosophy to promote polite drawing room conversation on scientific theory published just a few decades earlier. It's interesting to consider just what a young Michael Faraday would make of such a book or of *Harper's Electricity Book for Boys* (1907) in which the author, Joseph H. Adams, wrote:

> Theory is all very well, but there is nothing like mastering principles, and then applying them and working out results for one's self . . . The boy who makes a push button for his own home, or builds his own telephone line or wireless telegraph plant, or by his own ingenuity makes electricity run his mother's sewing machine and do other home work, has learned applications of theory which he will never forget. The new world which he will enter is a modern fairyland of science, for in the use of electricity he has added to himself the control of a powerful genie, a willing and most useful servant, who will do his errands or provide new playthings,

who will give him manual training and a vast increase in general knowledge.

True to his word, Adams's book touches only lightly on scientific theory, guiding the reader slowly through projects of increasing difficulty, including the building of several types of batteries. What turn-of-the-century boy could resist such an enticement to technological quest? To read Adams, building electrical devices was an adventure worthy of Jim Hawkins from *Treasure Island*, Huck Finn, and Tom Swift. For young boys of a certain age, inventor was added to the list of potential careers, along with the more banal and traditional choices of cowboy, pirate, and explorer.

Young boys weren't the only ones captivated by invention. For America's adult would-be inventors and tinkerers, who had been making do with scraps, Gernsback and others played the role of instrument maker by importing and distributing scientific and technical equipment from Europe that was not easily found in the United States. Those who had only been able to read about Marconi and other scientists finally had a reasonably priced opportunity to roll up their sleeves and explore the technology themselves in home workshops, though some of Gernsback's product line, such as the do-it-yourself x-ray unit, is actually frightening by today's standards.

It's worth noting that a variation of the Gernsback story was reprised in 1975 when a small Albuquerque, New Mexico, company, Micro Instrumentation and Telemetry Systems (MITS) introduced the Altair 8800 for $395 (about $1,500 in constant dollars) through the mail. Considered the first "microcomputer," the unit appeared on the cover of *Popular Electronics* in January 1975. The Altair (named after the brightest star in the Aquila constellation) offered no keyboard, monitor, or tape reader, and boasted a whopping 256 bytes of memory. Enthusiasts programmed it in binary machine language with toggle switches and little lights on the front panel. Thousands of the units were sold, and a Harvard student, William Henry Gates

III ("Bill" to his friends), contacted the company with an offer to write code for the machine.

HOWEVER, IT WAS GERNSBACK'S EVER-GROWING line of magazines in which he made his mark and allowed his imagination to really take flight. Starting out with a catalog to promote Electro Importing's products, he steadily included longer and longer articles. By 1908 Gernsback was a full-fledged magazine publisher with the launch of *Modern Electronics,* which also just happened to include a good many of Electro Importing Company's products along with instructions for do-it-yourself projects that made use of Electro Importing equipment. Something of a hybrid technical magazine and product catalog, it provided a printed forum for the technology enthusiast with stories on how to create home electronics, contests, and a "patent of the month."

Gernsback struck the perfect tone, offering more technical detail than the mainstream press without the narrow focus and stodginess of academic journals. There were other magazines catering to roughly the same market, but none with the breathless enthusiasm Gernsback managed to pack into his pages. Following his initial successes, Gernsback added more magazines to his offerings, including the world's first science fiction magazine, *Amazing Stories.*

Although often boasting lurid covers, Gernsback took pains with the content, reprinting classic stories by Jules Verne and others. His formula was simple: he put out the kind of magazine he would want to read. Gernsback is credited with coining the term "science fiction" within the pages of *Amazing Stories.*

Although now remembered for the Hugo Award, the science fiction award that bears his name, Gernsback was also something of an amateur inventor whose enthusiasm regularly exceeded his actual scientific acumen. His most notable invention, the isolator, was a helmetlike device that filtered out distractions to help people think and that he took to wearing around the office. Another device, the hypnobioscope, he claimed, could assist its users to learn while they slept. By the end of his life, he held more than eighty patents.

However, his true gift was for speculating on future technology, and at several points his predictions proved uncannily accurate. He imagined space flight, including multistage rockets, space walks, and radar along with a manned lunar landing by 1970—missing the Apollo 11 date by about a year.

What Gernsback either unleashed or promoted was the age of the radio hobbyist. Thousands of hobbyists around the country began building their own radio sets, first relying on units such as the Telimco, then improvising as they went along. Car batteries, spark

*Hugo Gernsback, the father of science fiction, got his start by selling a small wireless telegraph system to hobbyists. A scaled-down version of Marconi's famed technology, the Telimco Wireless Outfit was powered by two standard dry cells and boasted a limited range. However, it did not take long before hobbyists began tinkering with the basic package to boost its range with increased battery power and new types of antennae.*

coils, and even homemade batteries provided the power as they scavenged parts to build increasingly powerful transmitters.

A good many of these radio hobbyists would simply listen in on wireless transmissions from ships at sea, the news broadcasts regularly transmitted out to the ships, or they would try to make contact with other hobbyists. Some, however, like early hackers, took an aggressive approach, blocking transmissions from ships and news agencies with extended bursts from their homemade transmitters (called "brick on the key") or broadcasting false reports. In England, the Wireless Telegraphy Act of 1905 was designed to regulate all wireless communications, and even amateurs required a license to transmit or experiment. The United States had no such regulations because the occasional bills that came up in Congress were beaten down by the American Marconi Company and others lobbying against them. Very often a clean-cut teenage boy would be called to testify about his hobby and provide a statement against licensing.

The question was finally resolved when the *Titanic* sank in 1912 and hobbyists jammed the airwaves with false reports, prompting the first licensing of amateur radio operators in the United States. The sinking of the *Titanic* also led to one of early radio's more enduring legends. According to the tale, a young telegraph operator sat at his wireless set at the American Marconi station housed in New York City's Wanamaker's department store in April 1912 and received a seven-word message from the SS *Olympic*, "SS *Titanic* ran into iceberg. Sinking fast." The young man, again according to his own often-told account, tirelessly stayed at the set around the clock to receive regular updates.

The operator, David Sarnoff, the legendary pioneer in broadcasting, had begun as a newspaper boy and then moved to telegraphy with the Commercial Cable Company before finally landing at American Marconi. One of those bright young lads attracted to new technology, he ended up playing a pivotal role in broadcasting, though more than likely he created the *Titanic* story out of whole cloth. By the time the

*Titanic* sank, he had long advanced up the corporate ladder beyond the role of telegraph operator. What's more, as some historians have pointed out, the *Titanic* sank on a Sunday when the store, along with the American Marconi station, was closed. The *Titanic* story aside, Sarnoff did have a grasp of the potential for radio very early on, at one point authoring a lengthy memo that advocated "radio music boxes" for the home with broadcasts financed by the sale of radio sets—the hardware would pay for the software. When General Electric bought out American Marconi, changing its name to Radio Corporation of America (RCA), Sarnoff's dream of broadcasting and "radio music boxes" was realized.

EVEN AS GERNSBACK WAS MARKETING his battery-powered sets to eager hobbyists, they were already obsolete in terms of technology. Scientists in Europe, America, and beyond were actively searching for an alternative to the unwieldy coherer. While we may imagine that early wireless telegraphy conducted very much like the telegraphs connected by lengths of wire with operators tapping out messages at lightning speed, the truth of the matter was that sending even a simple message remained a long and arduous process. It took a lot of brute electrical force to broadcast a signal so that the coherer could do its work. The sparks were so large and deafening that operators took to wearing earplugs and the "key work" of sending the signal was a grueling task. To send a "dot" in Morse code the operator pressed down on the key for a full five seconds while dashes took more than fifteen seconds. A single short word frequently took more than a full minute to tap out.

There could be little doubt the coherer and decoherer needed to evolve. And evolve they did. As Gernsback and his wireless devotees enthusiastically sent their signals out into the ether, scientists around the world were independently searching for a viable coherer alternative.

Crystals had long been known to have some unusual properties when exposed to electrical current. In the 1800s Karl Ferdinand Braun made the discovery that electrical current seemed to flow more easily through some crystalline structures in one direction than in another. And, too, there was no doubt crystals now had some useful properties when it came to broadcasting. At Presidency College in Calcutta, Jagadish Chandra Bose filed for a U.S. patent for a crystal-based point-contact rectifier for detecting radio signals in 1901. In the United States, the AT&T engineer Greenleaf Whittier Pickard received a patent on a method for receiving radio signals that included a silicon point-contact diode in the summer of 1906. He would later go on to market crystal sets that used a "cat's whisker" of thin wire positioned against a crystal's surface to pick up the broadcasts without the use of batteries through a company called the Wireless Specialty Apparatus Company, which sold a unit called the Perikon—Perfect Pickard Contact.

Less than a year after Pickard received his patent, Henry Harrison Chase Dunwoody of the U.S. Army Signal Corps received a patent for a system using a point-contact detector made of carborundum (silicon carbide). Other patents were filed in Russia as well as Japan. And, in one of the stranger footnotes in technological and scientific history, Henry Joseph Round, who worked closely with Marconi, began experimenting with crystals with some unexpected results. Publishing his findings in the trade journal *Electrical World* (February 1907) he wrote,

> On applying a potential of 10 volts between two points on a crystal of carborundum, the crystal gave out a yellowish light. Only one or two specimens could be found which gave a bright glow on such a low voltage, but with 110 volts a large number could be found to glow. In some crystals only edges gave the light and others gave instead of a yellow light green, orange or blue. In all cases tested the glow appears to come from the negative pole, a bright blue-green spark appearing at the positive pole. In a single crystal, if

contact is made near the center with the negative pole, and the positive pole is put in contact at any other place, only one section of the crystal will glow and that the same section wherever the positive pole is placed.

He had unintentionally developed the first light-emitting diode (LED). A type of transistor, the LED's light is produced by the release of energy as electrons travel across the semiconductive material. For Round, the glow from the crystal was simply an interesting phenomenon, and not much came of his discovery until years later.

In the 1920s, Oleg Losev, a radio technician and self-taught scientist, independently discovered the same phenomenon in Russia. However, unlike Round, he continue his research into the strange glowing properties of crystals, publishing more than a dozen papers between 1924 and 1930 that largely went ignored by the scientific community. Losev, refusing to leave Leningrad ahead of the German siege during World War II, starved to death in 1942, and his work went unrecognized until recently.

However, the real breakthrough arrived in England, at University College London, when Sir John Ambrose Fleming came up with the idea of using a variation on the electric lightbulb to detect waves. He modified the bulb in such a way that it picked up the signals and converted them into electrical current. What he had invented was the first vacuum tube, which for years was popularly known as a "valve." Fleming, who had once worked for Edison, had the Ediswan Company make up the tubes, which he patented in 1904. They worked, though imperfectly.

FLEMING VACUUM TUBE

An American inventor, Lee De

Forest, improved on the concept, modifying Fleming's original design by adding a grid that could amplify the signal. He called this tube, which offered a marginal improvement over Fleming's design, the audion or triode vacuum tube. De Forest simply didn't have the scientific background to perfect the thing, and it wasn't until General Electric got involved with its teams of scientists and engineers that the vacuum tube became viable in 1912.

The amateur wireless operator with the telegraph key was also surpassed as early as 1900 when Reginald Fessenden wirelessly transmitted sound. A prodigy who graduated from college at fourteen, he was not a genius of the warm fuzzy variety. Arrogant, short-tempered, and impatient, the former schoolteacher somehow managed—despite lack of credentials—to hook up with Edison as a chemist for an extended period, but he left for a position at Purdue University and eventually ended up at a remote research station on Cobb Island in the Potomac that was meagerly funded by the U.S. Weather Bureau.

Fessenden ("Fessie" as Edison nicknamed him) used scavenged parts to build his radio, including an old cylinder from an Edison phonograph. The idea was to extend the traditional wireless telegraphic bursts into a continuous wave that could be modified by a voice. To receive the wave, he designed a replacement for the coherer, which he called a liquid barretter. This was something of a breakthrough device, consisting of a thin platinum wire immersed in an acidic solution that could receive a continuous signal. There was no need to tap it to reset a pile of metal filings.

Fessenden's first audio transmission traveled just one mile. The speech was distorted, but it worked, at least in principle. He kept at the work, trying to perfect it. Then, in early December 1906, he sent out a Morse code message announcing the first wireless transmission of speech on Christmas Eve. When Christmas Eve rolled around a few weeks later, at around 9:00 p.m., Fessenden tapped out a wireless message on a telegraph key—"CQ (seek you)"—to anyone who might be listening to his transmission from Brant Rock, Massachusetts. When a few ships at sea responded, he spoke into the microphone explaining that he was conducting a test of a voice transmission system, then picked up his violin and played *O Holy Night*, read a Bible verse, and signed off the air.

Standing nearby to witness the broadcast were representatives from General Electric and AT&T. The demonstration made an impression on the observers, but neither company liked the idea enough to fund it. Although impressive, the technology wasn't evolved enough to make it commercially viable. It wasn't until vacuum tubes could perform the work of transmission and receiving that radio actually became practical a few years later. Still, Fessenden had provided valuable proof of concept.

Not surprisingly, radio technology found its way into less than honorable professions—so much so that the magician Harry Houdini was prompted to write an article for *Popular Radio* in 1923 exposing the use of radio by phony spiritualists.

"Radio has given the 'spirit business' an enormous boost in the last few years," Houdini wrote. "While the rest of us have just been getting acquainted with it, many of the so-called psychics have been reaping a harvest."

Interestingly, Houdini got much of the basic technology wrong. In detailing the workings of a "talking tea kettle" and statues, he described not true radio transmissions, but a form of transmitting by way of induction. These "transmissions" emanate from the coil's electromagnetic field rather than bursts of Hertzian waves and are picked up at close range. First discovered in the 1830s and dubbed "electricity at a distance," the phenomenon had been discovered by Faraday, though he had little use for its commercial applications. Then, around 1887 Edison sent just such a wireless telegraph-like signal using induction coils but abandoned plans for its commercialization.

Some spiritualists also claimed that radio waves provided a link to the netherworld. The good Professor Lodge, who had so politely objected to Marconi's designation as the "inventor of wireless telegraphy," would later descend into the study of spiritualism and its relationship to Hertzian waves. However, most of the frauds were far more worldly. For instance, De Forest, who was desperately in need of funds to continue his research, was eventually lured into questionable Wall Street enterprises by stock promoters who wanted his name and credibility for stock schemes. At one point, he found himself quite literally encased in a glass laboratory on the roof of a hotel in downtown Manhattan.

ALTHOUGH RADIO, INCLUDING FESSENDEN'S DEVICE, did not find wide use during World War I, Marconi's wireless telegraphy played a vital role. In the first "modern war," all sides deployed transmitters and receivers, allowing for the coordination of troop movements, near real-time updates from the battlefield, and more efficient supply chains. The majority of the "portable" wireless sets used were enormous by today's standards. Some of these units weighed in at sixty pounds or more and required substantial antennas. "Portable" for World War I meant they were packaged for travel and could be comfortably transported by mule or caisson. Some sets offered hand-cranked dynamos (generators), though many still featured a full set of dry cell or liquid-based batteries.

Despite its still relatively primitive state during World War I, radio outperformed telephone communications on the battlefield as well as in the air. As *Popular Mechanics* reported in the spring of 1915, coded messages sent by wireless telegraphy were playing a key role in coordinating troop movements and locating targets.

> One of the first tasks of the wireless of the various warring countries was to fill in the gaps caused by severed cables. As later incidents have proved, however, that was only an insignificant portion of the work. Hundreds of miles of roaring battle line, hostile warships roving the remotest wastes of the sea, aeroplanes

> and Zeppelins soaring high above the earth, even the stealthy submarines, lurking in the depths for victims, are subservient to the invisible hand of the wireless.

As the magazine's story also noted, French and Belgian planes were equipped with hundred-pound radios capable of sending and receiving messages for more than fifty miles.

No doubt this was exciting stuff. The use of radios by armies and spies on both sides during the war piqued the interest of the amateur enthusiast who read breathless accounts of the technology in magazines like *Electrical Experimental, Popular Mechanics,* and *Popular Science.* It also pushed the radio industry forward, expanding the number of manufacturers and drawing additional engineers and scientists into the field.

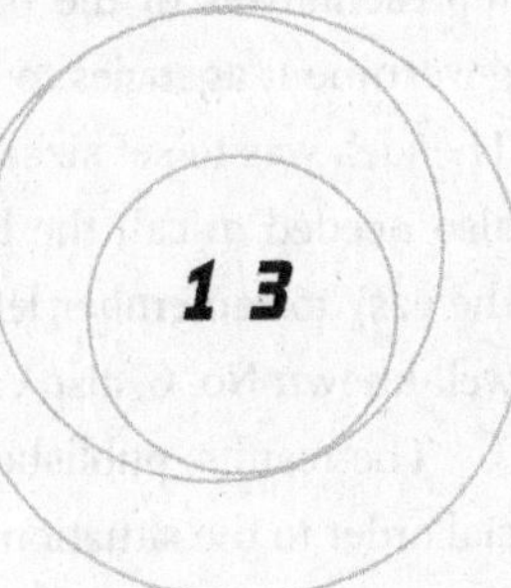

# 13

## *What Will They Think of Next?*

> *"Scientists take no needless risks! They take nothing for granted. Boastful or misleading statements are ignored, Entirely!"*
>
> *—magazine ad for Burgess Batteries, circa 1930*

The increased use of batteries and battery-powered devices in the battlefield did not go unnoticed by either industry or the military. Unfortunately, there were few standards for batteries. The common dry cell, known as the No. 6, most likely because it measured six inches high, was one of the few standard battery sizes available. All the rest were more or less ad hoc, assembled in small factories for specific devices, often by hand.

There was talk of standardization as early as 1912, but not much came of it until 1917 when the National Bureau of Standards (today the National Institute of Standards and Technology) met with representatives of the battery industry and the military and other government agencies to develop a set of specifications for batteries. The idea was to set sizes and minimum performance criteria. They also needed to call the batteries something, eventually settling on the easy to remember letters of the alphabet, except for the already well-known No. 6, also called a radio battery.

The results, published after the war in 1919, brought some official order to the situation for the first time. Toy and appliance manufacturers could now design products for specific sizes and voltages, and consumers could buy an electrical product with some assurance that batteries would be available far into the future.

The designations were modified over the years and more sizes added—the AAA, for instance—but now there were exact specifications for sizes and performance. Of course, unofficial standardization had been taking place for some time; radio manufacturers, for instance, designed the shelves inside early cabinet sets to hold specific-sized batteries.

BY THE LATE 1920S AND early 1930s consumer electronics had moved far beyond the novelties of light-up bowties and cane handles. Even flashlights, the most ordinary of electrical gadgets, had evolved. It was possible to buy pocket-sized sterling silver flashlights with fancy engraving or a woman's compact with a built-in light. And why not? After all, this was the age when accessories like gold lighters and cigarette cases—even the discreet hip flask—were common among the well-heeled.

With standardization, battery companies were pushed to become more competitive and creative. The problem they now faced was how to compete. How do you sell a standardized product? Well, fear works. In one early ad for Eveready, the headline blared: "I POURED A DEATH POTION FOR MY SICK BABY!" The ad then went on to recount in pictures and words how a Long Island housewife accidentally poured a dose of poisonous disinfectant rather than cough medicine in a darkened house. It was only after examining the bottle with a flashlight that she discovered her mistake. "Is it any wonder," she concluded her tale, "that I now write to let you know that my husband and I have *fresh DATED Eveready* batteries to thank for our baby's life?"

Another early ad by Eveready for its Daylo flashlights featured a fireman chastising a young couple, "You should have used a Daylo" while smoke poured from their house in the background.

At around the same time, Burgess Battery Company was taking a more scientific approach to marketing its batteries with an ad that announced, "No Store *'Around the Corner'* at the South Pole So the Byrd Expedition Couldn't Take a Chance." The ad went on to explain that "Scientists take no needless risks." What could be more credible than a battery approved by scientists? Actually, the founder, Charles Burgess, was a genuine scientist who founded the University of Wisconsin's Department of Chemistry and was instrumental in the success of Ray-O-Vac. Perhaps that's why Doc Brown in *Back to the Future III* chose Burgess No. 6 dry cell batteries to power the walkie-talkies he used to communicate with Marty McFly.

PORTABILITY WAS ALSO MAKING SOME advances, at least in niche products. For instance, it was possible to purchase a battery-powered electronic hearing aid. However, early models, like the Acousticon Model 28, produced in the 1920s, although battery-powered were as large as tabletop radios. The Acousticon was made by the Dictograph Products Company, which also manufactured office intercoms as well as one of the first eavesdropping devices, called the Detective Dictograph, also battery-powered. The Acousticon was intended to be set down on a table; a receiver was then fastened against the ear with a metal band. It was portable only in that there was a handle on top.

Although Ray-O-Vac claims to have invented the first wearable hearing aid in the early 1930s, other manufacturers had already produced wearable tubeless electronic devices that offered some amplification along with a great deal of distortion. It was the advent of vacuum tubes that made the devices practical though still not particularly portable, at least by today's standards. Batteries pumping out the three or six volts needed to fire up the tubes were bulky and carried in "holsters" under the arm or secured to the leg with cloth bands.

There were even some portable radios, sort of. In the early days, "portable" did not necessarily imply that transportation would be easy. A state-of-the-art line of portable radios made by a California company called Kemper Radio Lab in the 1920s featured the K-5-2 model that required ten batteries to power its five tubes and weighed well over twenty pounds. And then there's the RCA Victor Model P-31 (circa early 1930s), which was also considered portable. About the size of a small suitcase, the unit weighed in at over forty pounds. Radios were portable only in the sense that they were fitted into a sturdy case that included a handle.

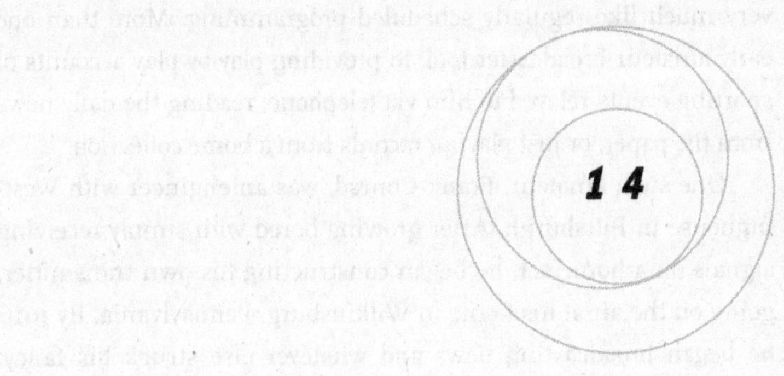


# Distance Dies in the Parlor

*"I am often asked how radio works. Well, you see, wire telegraphy is like a very long cat. You yank his tail in New York and he meows in Los Angeles. Do you understand this? Now, radio is exactly the same, except that there is no cat."*

*—attributed to Albert Einstein*

Just a few years after World War I, vacuum tube technology, if not perfected, was in a significantly better state than it had been before the hostilities. What's more, companies such as General Electric, Westinghouse, and RCA were investing in research programs and joint marketing agreements. Amateurs were still broadcasting from attics and garages, making up programming as they went along. These broadcasters, who had started out experimenting with the new technology, quickly began transitioning into something very much like regularly scheduled programming. More than one early amateur broadcaster took to providing play-by-play accounts of sporting events relayed to him via telephone, reading the daily news from the paper, or just playing records from a home collection.

One such amateur, Frank Conrad, was an engineer with Westinghouse in Pittsburgh. After growing bored with simply receiving signals on a home set, he began constructing his own transmitter, going on the air at his home in Wilkinsburg, Pennsylvania. By 1916 he began broadcasting news and whatever else struck his fancy, eventually adding music to his repertoire. After running through a supply of his own recordings, he joined forces with a local music store in exchange for mentioning their name on the air.

Within a short time Conrad was receiving letters, actual fan mail, and suggestions for his broadcasts. The popularity of his broadcasts wasn't lost on his bosses at Westinghouse, which was busily manufacturing sets of its own at the time. Soon plans were made to launch the first commercial radio station, primarily as a marketing tool to sell more radios. Based in Pittsburgh, KDKA went on the air on November 2, 1920. But was it the first commercial radio station? Some radio scholars give that honor to a smaller operation, WWJ, run by the *Detroit News*, which began broadcasting in August 1920.

Suddenly, there was something "on the radio" and by 1921 Sarnoff's once seemingly whimsical notion of a "radio music box" was very quickly becoming a reality. Professional radio stations grew at a pace that rivaled, perhaps even surpassed, the spread of the telegraph. Within just a few years hundreds of commercial stations went on the air throughout the country, and radio became the "must have" item for the home, taking up residence in the parlor or living room, displacing the piano as the family's primary source of entertainment. And in February 1922, President Warren G. Harding, who had been the first president to ride to his inauguration in a car, installed a radio at the White House.

. . .

RADIO WAS NOT ONLY A new technology, but a new form of entertainment, and broadcasters were making it up as they went along. When it was discovered that orchestras sounded better playing inside a tent, stations took to erecting tents inside studios. When the Tomb of the Unknown Soldier (now known as the Tomb of the Unknowns) at Arlington National Cemetery was commemorated on March 4, 1921, audiences listened to the broadcast speeches at Madison Square Garden in New York and the Auditorium in San Francisco.

THE BATTERY INDUSTRY CAUGHT ON early. Ray-O-Vac, based in Madison, Wisconsin, founded as the French Battery Company, began marketing dry cells for use in cars under the brand name Ray-O-Spark and branched out into D cell batteries for its own brand of somewhat comically named "French Flasher" flashlights. However, by 1920 the company was marketing a line of batteries for radios under the name Ray-O-Vac (for radios using vacuum tubes) with the help of the corporate cartoon mascot, Mr. Ray-O-Lite.

Radios had clearly moved beyond the hobbyists and were now mass produced by firms using many of the same assembly-line techniques adopted by the car industry. In 1922 sales of radios, batteries, and other accessories amounted to about $60 million though by 1929 that number rose some 1,400 percent to more than $800 million. Suddenly radio was big business. Of course Wall Street took notice and the words "radio" and "broadcasting" became something akin to what "dot com" was during the 1990s.

A company needed only the slightest connection to radio or broadcasting to see its stock price soar or attract investors. In 1928, Radio Corporation of America was selling at a low of 85 , though by 1929 at the height of the bubble it traded at more than 549. Of course, this couldn't last, and by late October of 1929 it was more or less over with the stock market crash.

. . .

THERE WAS VERY LITTLE UNIFORMITY of style to those early radios. Designers seemed to be struggling with the most basic concepts of what a radio should look like. Did form really need to follow function, or was something else required? And what kind of "user interfaces"—knobs and dials—worked best for the consumer? Some of the early sets looked like serious pieces of laboratory or industrial equipment while others were housed in polished wood cabinets, crafted to resemble fine pieces of furniture. What about the tubes and other inner workings? Should they be "black boxed" or put on display? And just how much should a radio cost?

The RCA-Westinghouse "Radiola Grand," with its gold-plated tuning dials sold for more than $300 back in the early 1920s (roughly $3,000 in today's dollars), while all manner of sets emerged from small manufacturing firms whose names are largely forgotten, such as A. H. Grebe & Co., American Auto & Radio Manufacturing, and Hi-Mu Radio Labs. If the price tag for one of the larger sets was too great, budget-minded consumers could buy one of the less expensive crystal sets and try their luck tickling the cat's whisker across a small galena crystal while listening in on headphones.

Companies like Ray-O-Vac and Eveready (National Carbon Company) got into the radio business with their own sets in much the same way they had entered the flashlight business. Eveready took it a step further, sponsoring a variety show broadcast called *The Eveready Hour,* with guests like Eddie Cantor and Will Rogers.

One company that stubbornly survived, Philco, grew out of the battery industry—its name was a contraction of Philadelphia Storage Battery Company. Founded in the 1880s as a supplier of batteries for arc lights, the firm turned to providing batteries to the doomed electric car industry before switching to radio batteries, and then, finally, to manufacturing home electronics.

. . .

THE VAST MAJORITY OF THE first radios were battery-powered. This made sense since the majority of American homes outside major urban centers were still without electricity. Private companies saw little profit in wiring sparsely populated regions. So radios were built to handle batteries, and many consumers took to using the bulky lead acid storage batteries out of their cars or farm vehicles to power the early sets. Those very early sets required not one, but three batteries. This was eventually cut down to two, called the A (which offered a low 1.5 volt supply) and B (which was significantly more powerful, pumping out ninety volts or more into the tubes). Not surprisingly, battery manufacturers loved this state of affairs, though their days as power suppliers for radios were numbered.

Stepping down current and smoothing out the pulsations of AC was a simple matter of adding a transformer and filter. That would eliminate the B battery. To eliminate the A battery meant the addition of a new tube, which essentially converted the pulsing AC current into a steady flow of DC. As early as 1926, manufacturers began building radios that could run off household current.

However, battery-operated radios were still popular in many regions. For instance, in rural areas, families whose homes offered neither central plumbing nor electricity quite literally listened to the radio by kerosene lamp. As late as 1935 only one American farm in nine had electricity, but a good many of them had radios to hear President Franklin Roosevelt's fireside chats and the farm reports.

As radio stations and radios in homes continued to multiply, news traveled into homes more quickly, and distance began dying for the average consumer. News of world events that had previously taken days or weeks to reach the general public could now be broadcast out almost immediately.

AS THE ROARING TWENTIES ROARED on, newspapers and media moguls of the day, including William Randolph Hearst and the *Los Angeles Times*, invested heavily in radio, seeking to get a foothold in the new medium. However, there were also more unlikely broadcasters entering the field. The Los Angeles–based faith healer Sister Aimee Semple McPherson, who had been offering prayer by phone for years—an early version of the crisis hotline—and enthusiastically employed stagecraft borrowed from the film industry, eagerly adopted the new technology. If thousands gathered at her Angelus Temple in Los Angeles's Silverlake district, she could reach tens of thousands more through radio. So it was almost no surprise when she became the first woman to hold an FCC broadcast license, cranking up the power to her own station KFSG (Kall Four Square Gospel) and America's first religious broadcasting station in 1924. More religious broadcasters followed, offering salvation, healing, and financial well-being over the airwaves, all for the price of a small donation.

Even as the broadcast industry was rapidly taking shape, a new generation of inventors was coming of age. Inspired by Gernsback's and other periodicals along with a growing number of technical books, young inventors, engineers, and scientists began to make their mark. One memorable inventor of the period, Samuel Ruben, would play an instrumental role in battery design. In his autobiography, *Necessity's Children: Memoirs of an Independent Inventor* (1990), Ruben, who was born in 1900, recalled a childhood of simple inventions from discarded items.

> A more troublesome problem, of course, was keeping myself stocked in materials for my various experiments. Mostly, I collected house discards of reusable materials—for example, candlewax-coated Quaker Oats cardboard containers for tuning coils—but I could not have gotten far without the assistance of an Italian junk dealer nearby. For a very small sum—usually what I could save from birthday coins my relatives gave me—he would sell me the items I needed, such as cotton-covered copper magnet wire and larger diameter bare copper wire. When, eventually, he asked me

> what I did with all these materials, I told him I was constructing a wireless telegraph set. He looked up and beaming with Italian pride, exclaimed, "Ah Marconi!" After that, he would give me small items free and would inquire about my wireless experiments.

Ruben's account of his childhood experiments in the early part of the twentieth century is fascinating for his ingenuity and determination. Although he never completed his formal education beyond high school, he managed to master the sophisticated technology of the day through experimentation and reading. Even at the end of his career, he fondly recalled reading the works of Faraday. "One could do worse than to come under the influence of such a man and such a mind at the age of fourteen," he wrote.

Ruben looked back fondly too on the Electro Importing Company and Gernsback's publications. "Later that same year, 1914, I came across an announcement for a contest in a magazine that dealt with amateur radio and other electrical equipment," Ruben recalled.

> The magazine was published by a company called Electro-Importing Company [*sic*] and edited by a man named Hugo Gernsback. The company had a store on West Fulton Street that sold complete radio receiving and transmitting sets, along with components parts for amateurs. The contest in question offered a one year's subscription as a prize for the design of a portable optical signaling device. I submitted a sketch for such a device: A wooden camera tripod supporting a board on which was mounted a flashlight whose switch was connected to a telegraph key. My submission won the prize, and my drawing, along with the details, was published in the magazine.

It is telling that a man who held more than 300 patents at the end of his career still remembered with pride the prize and notoriety Gernsback bestowed on his first invention.

Among Ruben's inventions was a device that he called a "trickle charger" to keep batteries charged using household current; then in 1926 he came up with the "battery eliminator," an aftermarket device that allowed radios to plug into wall sockets. Combined with the widening power grid that was slowly making its way across the country, battery power was becoming obsolete as the primary power source for new consumer technology. In the end battery-powered radios took on the somewhat disparaging nickname "farm radios."

LATER, RUBEN WOULD GO ON to pioneer the alkaline manganese battery in the 1950s in the new AAA size for Kodak's line of flash cameras. He was a new type of inventor, who, unlike Edison, did not manufacture or sell the devices he created. By all accounts, he had no aspirations of building a business empire of factories and offices. Rather, he licensed his inventions to firms that already had the infrastructure to bring them to market.

One of those companies, P. R. Mallory, was a manufacturer of tungsten filaments for lightbulbs and a limited line of switches, founded by Philip Rogers Mallory. The heir to the Mallory Line, a coastal shipping line dating back to the 1860s, Mallory more or less left the family business to pursue technology, specifically electrical technology. And, it was with Mallory that Ruben found something close to the ideal business relationship. Mallory, although a pragmatic businessman, also had a soft spot for inventors. Unlike many of the technology firms of his time as well as today, he was not averse to licensing products developed outside his company's own R&D efforts.

P. R. Mallory would eventually shift entirely into the battery field and then, following Mallory's death, move through several owners, including Dart Industries, Kraft Foods, Wall Street investors, and, finally, Gillette, along the way changing its name to Duracell.

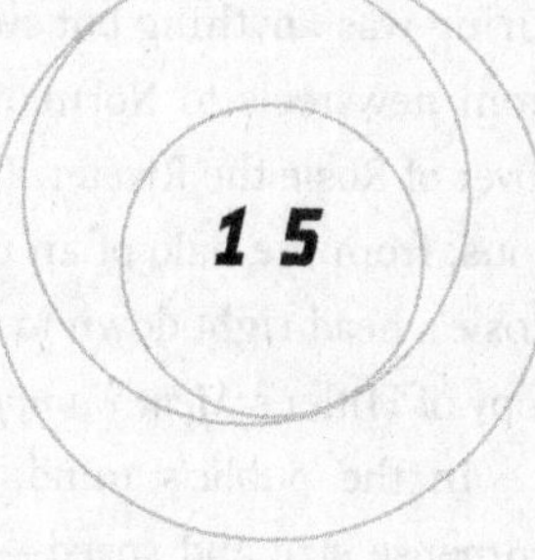

# The Endless Frontier

*"A spider web of metal, sealed in a thin glass container, a wire heated to brilliant glow, in short, the thermionic tube of radio sets . . . Its gossamer parts, the precise location and alignment involved in its construction, would have occupied a master craftsman of the guild for months; now it is built for thirty cents. The world has arrived at an age of cheap complex devices of great reliability; and something is bound to come of it."*

*—Dr. Vannevar Bush, Director of the Office of Scientific Research and Development (OSRD)*

Bombers flowing off assembly lines and warships splashing into the water after christening are iconic images of America's World War II industrial effort. Sources of pride and propaganda, reports from factory floors on the home front were nearly as ubiquitous and dramatic as dispatches from the distant battlefields of Europe and Asia. The portrayal of American manufacturing was anything but subtle and was documented in everything from newsreels to Norman Rockwell's 1943 *Saturday Evening Post* cover of Rosie the Riveter. The message was unashamedly unambiguous, from the halo of an upturned protective welder's mask above Rosie's head right down to her foot resting comfortably on a ragged copy of Hitler's *Mein Kampf.*

In the public's mind, America's manufacturing capacity—its immense size and speed—became linked with military superiority and inevitable victory. "So the American way of war is bound to be like the American way of life . . . It is the army of a nation of colossal business enterprises, often wastefully run in detail, but winning by their mere scale and by their ability to wait until that scale tells," wrote the Cambridge professor D. W. Brogan in *Harper's Magazine* (May 1944).

Less well known is the parallel, though no less ambitious, scientific and technological war effort. The Manhattan Project, of course, has been well documented, but there were other facilities and efforts that have fallen into obscurity. The Rad Lab at MIT, which worked to perfect the British system of radio detection and ranging (radar) worked on the cutting edge of science and employed some 3,000 personnel to create some hundred different radar systems by the end of the war.

College professors, along with bright graduate and undergraduate students in engineering and talented hobbyists, were pressed into service. At one point during the war, Bell Labs alone employed nearly a thousand scientists and engineers working solely on military projects. Personnel and facilities expanded so quickly that it was not uncommon to see temporary Quonset huts and hastily constructed wooden structures popping up almost overnight adjacent to corporate headquarters. Most of this work, of course, was done in secret, hidden away from the newsreel cameras. And Rosie, unfortunately, had no iconic equivalent among the lab rats who labored to create the next generations of weaponry.

· ·

IT WAS ONLY AFTER THE war ended, as books and magazine stories began to appear, that the full scope of the enormous wartime scientific effort began to emerge. In addition to large weapons systems, World War II was a war fought with sophisticated portable devices, many of them relying on batteries.

The most intriguing product of this scientific push was the proximity fuse. Although largely forgotten now, except by military history buffs, the secrecy and manpower that went into creating the technology was surpassed only by the Manhattan Project. Classified as top secret during the war and largely forgotten today, the proximity fuse pushed the limits of the technology in an effort that required scientists and engineers scattered across a half dozen or more facilities with Johns Hopkins's Applied Physics Lab (APL) serving as the hub. With research begun in 1939 by British scientists, the effort was taken up in the United States by the wartime National Defense Research Committee (NDRC) and its Office of Scientific Research and Development (OSRD) in 1940. From there it grew to include a multitude of military, civilian government, and private entities.

IN THEORY THE PROXIMITY FUSE was an easily understood device based on perfected technology. The idea was simple: a fragmentation bomb that exploded near a target—say within a hundred feet—would be more effective in combat than a projectile or traditional bomb that needed a direct hit. The comparison was one between a shotgun and a rifle. What's more, the weapon was necessary, if only for antiaircraft artillery, since planes had become faster and more durable in the years between the two wars.

What was needed was a projectile with an electronics system; radio technology would provide the key. A small, basic transmitter that bounced a signal off a plane could trigger the detonation by way of an equally basic receiver capable of picking up those returning radio waves. If engineered right, such a weapon could even blow the German V rocket bombs out of the sky before they reached London.

Simple, except for the design parameters. Engineers could easily make a "bench model" that would work just fine, as long as it stayed safely on the bench. In the field, the unit had to withstand the massive G-Force—something like 20,000 Gs—impact on firing and then even more Gs with the natural spin as it cruised through the air. It's fair to say that no electrical device had ever been designed for that kind of punishment. The unit had also to be small enough to fit on top of standard-size artillery shells, which meant pushing the limits of miniaturization.

To do this, the teams designed extraordinarily small subminiature vacuum tubes, each tube just slightly larger than a pencil eraser. Very early on, it became obvious that standard wiring wouldn't work. Radio wiring and all that went along with it was good enough for the living room radio, but far too large to mount on an artillery shell.

To solve the problem, the engineers perfected the concept of the circuit board or printed circuits. Developed in the mid-1930s by the German refugee Paul Eisler, the process used a conductive foil rather than wires to make connections between different components. Although others had been working on similar processes as a means to reduce the jumble of wires in telephone and telegraph switching stations and reduce factory mistakes in wiring radios, Eisler seems to have brought the system up to date.

And then there was the battery. The battery had to be reliable for several seconds, but also capable of being stored for years without losing its charge. National Carbon came up with what may be one of the greatest design solutions in battery history. The battery looked very much like a miniature voltaic pile. However, the small metallic disks, one stacked on top of the other, had a hollow center into which was placed a glass ampoule of electrolyte. When the shell was fired, the ampoule shattered and the natural spin of the projectile distributed the fluid, activating the battery in midflight. From an engineering standpoint, it was a brilliant solution, incorporating the extreme design parameters into the battery's function.

In truth, this was not an entirely new concept. A similar technical solution was used in the Hertz horn or chemical horn naval mines developed in the 1860s by the Prussian scientist Dr. Albert Hertz. In Hertz's design, which was later perfected by other countries, the battery to detonate the mine was activated when a glass vial filled with acidic electrolyte was broken by a passing ship's hull.

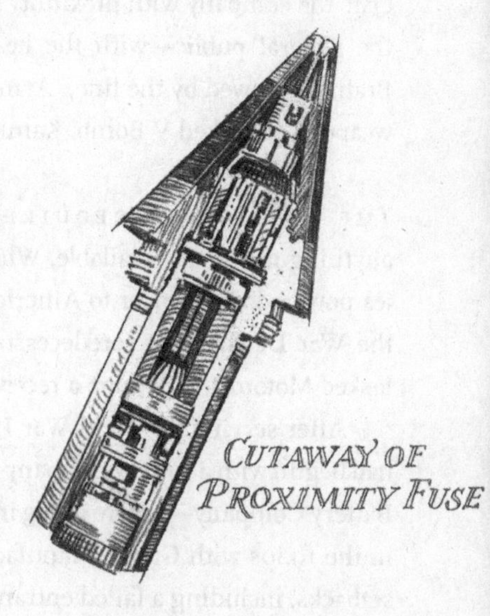
CUTAWAY OF PROXIMITY FUSE

Still, what made the proximity fuse battery so clever was the fact that the engineers had taken an outdated wet battery format, updated it slightly, and then incorporated what were seemingly insurmountable obstacles—the G-force and rotation—into the design. It was very much a case of engineering lemonade from some very large technical lemons.

The fuses were a breakthrough, but that also proved a problem. What if the enemy got hold of one? If a dud landed harmlessly on the battlefield, the enemy could reverse engineer the thing to use against American targets. Playing it safe, the fuse was first cleared for fighter plane combat in the Pacific theater in 1943. If there was a dud, it would splash into the ocean and sink. It wasn't until 1945, toward the very end of the war, that the fuse saw widespread use during the pre-invasion bombardment of the Battle of Iwo Jima. In the end, millions were produced and fitted into numerous types of artillery armaments.

After the war, a 1946 magazine ad for Eveready sought to associate the company with proximity fuses—virtually unknown among the general public—with the headline, "The Shell with a 'Radio Brain'" followed by the line, "Army, Navy lift censorship on mystery weapon that licked V Bomb, Kamikaze Attacks."

THE WAR EFFORT REQUIRED RELIABLE power far beyond anything previously available. What changed was what these batteries powered. Even prior to America's entrance into the war in 1941, the War Department (predecessor to the Department of Defense) tasked Motorola to develop a receiver-transmitter for the front lines.

After serving in World War I, Paul Galvin, Motorola's founder, had begun with a battery company—the Chicago-based D&G Storage Battery Company—before going into the battery eliminator business in the 1920s with Galvin Manufacturing Corporation. After several setbacks, including a failed entrance into the home radio market, he entered the car radio market under the name Motorola.

*World War II was the first "battery-powered war" with the introduction of portable radios into combat zones. Back on the home front, factories ran around the clock to produce enough batteries to power the war effort.*

Now, with a government contract in hand, he produced a portable radio, the Radio Set SCR-300. It could be transported into combat zones by a single soldier, but not particularly easily. Weighing in at over thirty pounds, with a range of about three miles, the unit was rigged as a backpack and featured a telephone-type handset. Motorola produced some 50,000 of the units for the war. Sealed in a steel case, the SCR-300 required eighteen vacuum tubes powered by a single, unique shoebox-size B-80 battery that provided three different power levels for different circuits.

Designed specifically for foot soldiers in the vanguard, the more portable SCR-536 handheld walkie-talkie (or Handie-Talkie) followed in 1941; looking very much like an enlarged version of those clunky cell phones from the 1980s, the device had a range of about one mile and required five tubes and more than a pound of batteries to fire it up. The thing weighed five pounds, and typical battery life was about a single day. Still, it incorporated a few interesting design features. The bulk telephone handset was gone. The case itself functioned

*Early battery advertisements often featured action-packed adventures in which the always reliable battery was the hero of the story.*

as a handset—similar to today's cell phones—with the mouth ear-pieces positioned at the top and bottom; the on/off switch was activated when you pulled out the giant forty-inch antenna. Another key advantage was that these were crystal-controlled sets—the crystals being quartz—so that tuning was done by changing out a crystal and not fiddling with the dial. By the end of the war, Galvin had manufactured more than 100,000 of the units.

The question remains: just who invented the walkie-talkie? Like the backpack, it was manufactured by Galvin. However, an independent inventor by the name of Al Gross is also credited. A ham radio operator from Cleveland, Ohio (call sign W8PAL-Gross), is said to have thought up the idea as early as 1938. Fascinated with radios since he sneaked into the radio operator's "shack" on a cruise with his parents when he was nine years old, Gross came up with the idea of a small, handheld radio while still a teen.

And this is where the story gets complicated. Recruited by the Office of Strategic Services or OSS (forerunner to the Central Intelligence Agency) during the war, Gross worked on what would become known as the Joan-Eleanor project. The small, four-pound, handhelds known as Joans (named for a WAVE Major, Joan Marshall), were issued to intelligence agents behind enemy lines. Compact, they weighed just four pounds and measured 6.5 inches by 2.25 inches by 1.5 inches. The Eleanors (named for Eleanor Goddard, the wife of one of the engineers on the project) weighed some forty pounds and were mounted in aircraft flying at 40,000 feet.

Adding to the confusion were the efforts of the Canadian Donald Hings, who is said to have invented a portable two-way radio in 1937 while working for Consolidated Mining and Smelting Company in Vancouver.

In the end, it was Galvin who received the patent for the walkie-talkie, since patents couldn't be issued for secret spy gear. And, for those given to quibbling, a case can be made that technically, Gross had only invented half of a walkie-talkie, since half of the system was mounted in an aircraft.

THE MINE DETECTOR WAS INVENTED by a Polish military engineer named Jozef Kosacki who was living in England after the German invasion of Poland in 1939. It was while working at St. Andrews in Scotland that he came up with a device using available technology—a long pole with a flat disk on the end holding two coils in parallel. Weighing in at under 30 pounds, one coil at the end of the pole sent out an oscillating signal and the other received it while the operator listened in on what was essentially a telephone strapped to his waist with a headset. When a metallic object, such as a mine, interrupted the signal, the operator could clearly hear it. Kosacki never patented the device, called the Mine Detector (Polish) Mark I or simply, the Polish Detector, which no doubt saved thousands of lives, giving it to both the British and Americans.

There were weapons as well, such as the M1 rocket launcher, which fired small fin-stabilized rockets against armored vehicles and emplacements. Fired from the shoulder, the M1 launcher was quickly nicknamed "bazooka" by troops in the field after the nonsensical instrument played by radio comic Bob Burns, which he improvised from plumbing pipes. The original bazooka rocket was ignited by two standard-size D cells in the stock, though later models eliminated the batteries altogether in favor of a small magneto activated by the trigger.

Innovative batteries were also introduced at sea. By the end of the war, the U.S. Navy had developed the MK26 torpedo, which used seawater as an electrolyte. This would not only reduce the weight of the torpedo, always a factor aboard ships, but also provide virtually unlimited storage time, since the batteries never "went bad." Though the MK26 never saw combat, the battery system, developed by Bell Telephone, did provide proof of concept for later torpedoes. The development of the seawater battery would also lead to more peaceful applications, such as automatic triggering of rescue beacons.

Batteries powered the first signal beacons for downed pilots, a forerunner to both the identification transponders and black boxes now carried on private and commercial aircraft. Large bulky things, they consisted of stacks of reinforced batteries and crudely "shock-proofed" circuitry powering a simple radio transmitter in a sturdy metal frame that automatically sent out an SOS signal when the plane crashed.

Naturally, there were also innovations on the other side, notably the Enigma machine. Despite popular belief, the Nazis didn't invent the Enigma; it was originally built in the 1920s and intended for businesses. As one brochure printed in English proclaimed, "One secret, well-protected, may pay for the whole cost of the machine . . ." Scaled down considerably over the years, the device was not only portable enough to carry around easily, but featured a series of small electric "lamps" that were battery powered.

A MAJOR PROBLEM WITH MUCH of the new electronics going into the battlefield wasn't the engineering, but the batteries. Very early on, the military discovered that batteries do not do well in the tropics. Batteries shipped to the Pacific and North African theaters of war arrived depleted. It seems the heat and humidity was speeding up the chemical reactions. What was needed was a battery that could function in any environment.

Sam Ruben came up with the solution. Working with new chemistries and containers in his New Rochelle, New York, lab, he hit on the mercuric-oxide or mercury cell, the first new battery chemistry in over a century. The battery worked well, but Ruben, who had only a tiny lab, couldn't produce the millions of batteries the war effort required and handed off the contract to P. R. Mallory Company (later Duracell). With its works classified as "top secret," the company turned out millions of the batteries, later known as the Ruben-Mallory or RM cells, to power the war effort with the company running round-the-clock shifts to meet demand for the "sealed in steel" batteries that powered everything from field radios to the newly designed L-shaped flashlights that soldiers could wear on their belts. Later, as the war intensified, Ray-O-Vac was brought on board to help fill the orders.

As for Ruben, when informed that his royalties would total some $2 million a year, he promptly renegotiated the terms downward. "I felt that it would be unconscionable to receive such large royalties for military requirements during wartime," he later wrote. "Consequently, we agreed upon a payment of $150,000 per year, which would amply cover laboratory operation and staff expenses."

And, in one of the stranger technological footnotes, Ruben saw the unlikeliest use for his batteries appear in 1957 when the Soviet Union launched Sputnik (Russian for co-traveler), the first man-made satellite. The twenty-three-inch diameter satellite that circled for three weeks contained not only a radio transmitter, sending out little more than beeps or pulses, but also a temperature regulation system. Surprisingly, both were powered by batteries suspiciously similar to Ruben's design, at least according to a story in the Soviet publication *Young Technique* in October 1957. Apparently, the American military had shared the top-secret technology with the Soviet Union during the war. Even more surprising, in 1961 an official Soviet government publication called *Knowledge Is Power* gave credit to the United States for development of the batteries.

It would not be the last time Ruben's batteries traveled into space. During the failed Apollo 13 mission, astronauts used light-up pens powered by Ruben's cells as a light source.

> "As you know, due to the explosion, we were forced to ration our electric power and water. With regards to the former, we never turned on the lights in the spacecraft after the accident," the astronauts James Lovell, Fred Haise, and John Swigert wrote Ruben. "As a result, your pen lights served as a means of 'seeing' to do the job during the many hours of darkness when the sunlight was

> not coming through the windows. We never wore out even one set during the trip; in fact, they still illuminate today . . . Their size was also a convenience, as it was handy to grip the lights between clinched [*sic*] teeth to copy the lengthy procedures that were voiced up from earth."

America emerged victorious from World War II with companies fattened by profits from "cost-plus" wartime production and endowed with new technology and processes. The infrastructure and knowledge base so hastily assembled during the war was now quickly put to peacetime use. Not only was America's physical infrastructure of factories, rail lines, and talent pool intact, there was another key component, the GI Bill (Servicemen's Readjustment Act of 1944). Fearing the economic impact of millions of American soldiers returning home en masse, FDR signed the bill nearly a year before the German surrender.

In addition to providing low-interest, no-money-down mortgages, it also offered college tuition and student stipends for returning vets. By 1947 close to half of the 16 million war veterans were either enrolled in college or receiving job training. At one point veterans made up nearly half of the college students in the United States. All told, some 91,000 scientists and 450,000 engineers studied through GI Bill benefits following World War II, including 14 Nobel Prize–winners in science.

Some saw the peacetime potential of the advanced technology early on. Dr. Vannevar Bush, who envisioned and then headed the National Defense Research Committee as well as its 2.0 wartime version, the Office of Scientific Research and Development, charged with applying the latest technology to warfare, was quick to spot the future. In two landmark essays, "As We May Think" (*The Atlantic*), and "Science the Endless Frontier: A Report to the President," he exhibited uncanny prescience as to the future role of technology. Both pieces were written in the summer of 1945, months prior to Japan's September surrender that marked the end of the war.

Bush, who had received his doctorate from MIT, was something of an amateur inventor himself as well as the cofounder of the American Appliance Company, which would eventually morph into Raytheon (Greek for "light from the gods"), a major defense contractor.

He was that rare member of his nineteenth and twentieth century–spanning generation who was not only welcoming of technological change, but could also extrapolate with a fair amount of accuracy the role it would play in society. Though born in the noisy coal- and steam-powered nineteenth century when most Americans lived on farms, Bush was able to foresee an inevitable future of increasingly advanced circuitry, more sophisticated devices, and an expanding role of technology emerging from cutting-edge science. In "As We May Think," he imagined a device he called a "memex" that many have compared to the Internet, though it more closely resembled an enormous database.

However, it was in "Science the Endless Frontier" that he forcefully advocated a national effort to promote science. "The pioneer spirit is still vigorous within this nation," Bush wrote to FDR. "Science offers a largely unexplored hinterland for the pioneer who has the tools for his task. The rewards of such exploration both for the Nation and the individual are great. Scientific progress is one essential key to our security as a nation, to our better health, to more jobs, to a higher standard of living, and to our cultural progress."

What happened after the war, as Bush's predictions proved more or less accurate, was something of a mid-twentieth-century version of "beating swords into ploughshares," that is to say, rewiring bombs into radios and televisions. The value of military technology was not so much in the armaments themselves, but in their components and the processes used to create them. Printed circuit technology, for example, perfected for proximity fuses, seemed to hold particular fascination for private companies.

In a postwar publication issued by the National Bureau of Standards, Cledo Brunetti and Roger W. Curtis reported,

> . . . printed circuits are now the subject of intense interest of manufacturers and research laboratories in this country and abroad. From February to June 1947, the Bureau received over one hundred inquiries from manufacturers seeking to apply printed circuit or printed circuit techniques to the production of electronic items. Proposed applications include radios, hearing aids, television sets, electronic measuring and control equipment, personal radiotelephones [*sic*], radar, and countless other devices.

By using the printed circuit board technique, in which connections between components were essentially painted on the board, engineers could eliminate much of the birds-nest wiring and the galvanized chassis common in many electrical devices. They could also cut production costs and more or less reduce the wiring of a device to two dimensions and fit a good deal more circuitry into a compact space.

In old movies and television shows, a standard joke was to hit or shake a broken electrical appliance to get it to work. This method was actually often effective—at least temporarily—to reestablish a contact between loose connections. Fix-it shops and television repairmen, both long gone from the American landscape, did a thriving business by hunting out and resoldering loose connections. With the advent of the circuit boards, their days were numbered.

Industry had learned a few other tricks from wartime requirements, like effectively using multipurpose vacuum tubes so that receivers needed only four or five tubes instead of the standard seven or eight. Engineers also began placing components closer together to save space while some simple components even performed double duty. For instance, the World War II–era walkie-talkies were turned on by pulling up the antenna. Why not apply the same principle to consumer and industrial products?

A good many of these clever design innovations came directly from combat requirements, particularly efforts in reducing size. In a lengthy monograph on miniaturization and microminiaturization following the war that was produced for the U.S. Army Electronics Command at Fort Monmouth, New Jersey, the command historian, William R. Stevenson, wrote, "Miniaturization as a major design goal in communications-electronic equipment began to be felt only after the requirements of the Armed Forces began to be impressed on industry. Here adequacy for combat service in many cases was so dependent on size, that miniaturization had to be employed regardless of cost."

This was all good news, of course, except for the batteries. Although relatively cheap, reliable, and long-lasting, they still weren't up to the task of powering more than a few tubes, even subminiature tubes, for any length of time. No matter how small you made vacuum tubes—and companies like Raytheon could make them very small indeed—they were still power-hungry little beasts.

Quite simply, the battery industry had run into the brick wall known as Faraday's First Law of Electrolysis, which logically states that in order to double the output of any battery, the amount of material in that battery must be doubled. New materials, such as mercury and cadmium, had given batteries longer life and more power, but the output was still far too low or the materials much too expensive to meet the needs of the American consumer for any kind of really sophisticated device. And if you had to wait a minute or so for the tubes to "heat up," the inconvenience was minimal. Even when tubes "blew out," which they did with annoying regularity, it was easy enough to pull the most likely culprits and test them on countertop devices at the local hardware and drugstores that also stocked replacements. And if radios and other appliances weren't particularly portable, then that was fine, too. That's just the way things were.

DESPITE THE PROBLEMS, SOME NICHE companies did strive for miniaturization. Hearing aid manufacturers, for instance, made use of subminiature tubes. Raytheon sought to establish itself in the consumer market, bought up a Chicago company called Belmont Radio Corporation (another wartime manufacturer transitioning back into peacetime products) and quickly introduced what is widely believed to be the first "pocket radio." Measuring just 3 inches wide, 6 inches high, and less than an inch thick, the Belmont Boulevard required three batteries, a 22.5 volt B and two 1.5 volt A batteries, to power its five subminiature tubes.

Did people want a miniature and portable radio with an earpiece instead of a speaker? Apparently they didn't, at least not Belmont's version. The Boulevard (model 5P113) was a failure in the marketplace, though today it's extraordinarily rare and much sought after by collectors.

However, what's really interesting about the Boulevard is just how little it resembles a "pocket radio" or any of today's portable electronic devices. The metallic case was a two-tone gold and silver, and customers had a choice of trim that included Moroccan leather, pin seal, alligator, or suede. Clearly the Boulevard was a rich man's toy or gentleman's accessory, very much like a solid gold cigarette case, engraved hip flask, or sterling silver flashlight. From the twenty-first-century perspective, the Boulevard looked very much like the kind of personal possession designed to become an heirloom, though it was packed with technology destined for obsolescence.

Another company that saw the potential in miniaturization and battery-powered products was the Hamilton Watch Company. A well-known American manufacturer of timepieces with a history that stretched back more than a century, Hamilton made a very large bet on battery power in 1946 by launching an ambitious effort to produce an electric wristwatch. Hamilton probably didn't really know what it faced, since it took more than a decade to bring the watch to market.

Not only did the company have to develop the watch itself, but Hamilton started off by also trying to design a new kind of battery to power it. In the end, the battery proved too much for the watchmakers and they brought in National Carbon (not yet Eveready) to modify one of Sam Ruben's button battery designs to create a 1.5 volt battery called the energizer.

Still, the firm was optimistic. The effort moved ahead slowly year after year. Then, a lengthy 1956 story in the *New York Times*, months prior to the release of the watch, outlined Hamilton's plan:

> "A revolution is coming to the ancient art of timekeeping, specifically in the design of watches," the story enthused. "It is being sparked by new discoveries in electronics and advances in miniaturization (the process of getting more and more equipment in less and less space), and it will take two forms. Scheduled for the immediate future is the electric watch, no bigger than the one you are wearing, which will run entirely on the current from a battery. In the more distant future, say 1975, is an 'atomic' wrist watch, a time piece operated by a midget nuclear power plant."

Finally, in early January 1957, Hamilton announced to the world the first electric watch. Called the Hamilton 500 or Hamilton Electric, the watch retailed for $175.00 (around $1,300 in constant dollars) in solid gold and $90.00 ($700 in constant dollars) for gold-filled. Unfortunately, despite more than a decade of work, the technological design left much to be desired. The $1.75 energizer battery tended to discharge more quickly than the year-long life cycle Hamilton had envisioned. And, too, the watch was something of a quirky hybrid. Rather than redesign the works, Hamilton simply replaced the mainspring with a small motor-type arrangement to power a balance wheel and traditional array of gears, which didn't make for the most reliable timepiece.

Still, it was the "watch of the future." The 1950s were, after all, a time of unbridled and quaintly fanciful predictions when it came to future technology. At a time of unprecedented prosperity, the possibility of personal flying cars, robotic housekeepers, and vacations on Mars seemed more than likely at some future date. Disneyland opened one of its more popular attractions, Tomorrowland, which included a TWA Moonliner rocket ship. And why not? The future along with a whole world of technological miracles was arriving very quickly indeed.

Press reports marveled at the size of the electronic watch's battery along with the *newness* of the concept. The little battery was enthusiastically described as able to rest on a fingertip, as large around as a shirt button or an aspirin, and often photographed alongside the watch itself. Since the watch looked rather ordinary, the little energizer was the star.

Perhaps because the new electric watch looked so much like a standard Hamilton—nobody could tell you were wearing a piece of the future on your wrist—the company released a model in a futuristic asymmetrical case. Elvis bought one and so did the host of the *Twilight Zone* television show, Rod Serling. Even more recently, Will Smith and Tommy Lee Jones wore a pair of Hamilton Electrics for their roles in the *Men in Black* movies.

Production on the Hamilton Electric ended in 1969 with little fanfare as quartz watches began arriving on the marketplace, beginning with a very limited production run of the Seiko 35 SQ, which sold for about $1,200 (more than $6,000 in constant dollars). The Hamilton Electric's life as a consumer product lasted just about as long as the time it took to develop it. The "atomic wristwatch" never came to pass, though the National Bureau of Standards did produce an atomic clock as early as 1949, and today consumers have a wide

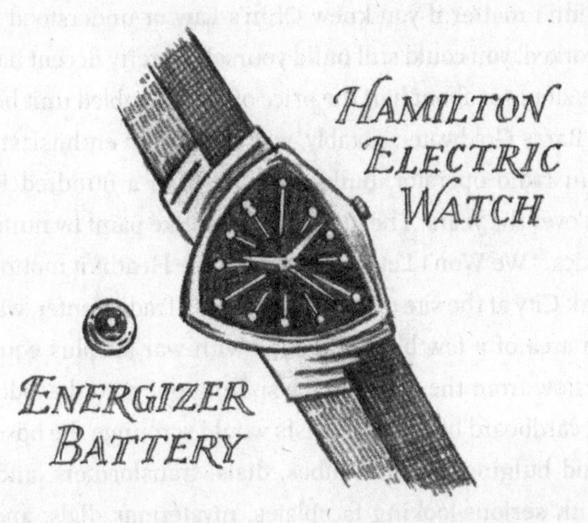


choice of timepieces capable of picking up its broadcast signal for "atomic accuracy."

THE AMERICAN CONSUMER HAD DEVELOPED a taste, even a fascination, for electronic gadgets following the war. Virtually every household in America was wired for electricity, thanks in large part to the Rural Electrification Act of 1936. Even as batteries were becoming marginalized to power toys, flashlights, hearing aids, and a few simple devices, consumers and hobbyists could not get enough of electronic gadgets.

In 1947, a small company in Michigan began buying war surplus components to repackage them into do-it-yourself kits beginning with oscilloscopes, then moving on to more consumer-oriented devices such as ham radios and phonograph amplifiers. The Heath Company became Heathkit, a mail order outfit, not unlike Hugo Gernsback's Electro Importing Company. The gimmick was that do-it-yourself hobbyists needed only a couple of basic tools to complete the assembly. A few "pleasant evenings at home," the catalog copy promised, was all it took to build a state-of-the-art amplifier.

It didn't matter if you knew Ohm's Law or understood how the thing worked, you could still build yourself a pretty decent ham radio or home stereo at about half the price of an assembled unit bought in a store. Barry Goldwater, notably, was a Heathkit enthusiast, as well as a ham radio operator, building more than a hundred Heathkit projects over the years. The idea was not unlike paint by numbers for electronics. "We Won't Let You Fail," was the Heathkit motto. And in New York City at the site of the future World Trade Center, was Radio Row, an area of a few blocks flooded with war surplus equipment, the overflow from the dusty stores spilling out onto the sidewalk in bulging cardboard boxes. Hobbyists would scrounge the boxes brimming and bulging with old tubes, dials, transformers, and equipment with serious-looking faceplates, mysterious dials, and toggle switches. Newspapermen, tow truck operators, and ambulance chasers made pilgrimages down to Radio Row to snatch up army surplus tank radios to monitor police and fire calls.

THEN THE ENTIRE GAME CHANGED, though deceptively quietly, in 1947 when two Bell Lab scientists, John Bardeen and Walter Brattain, in their fourth-floor lab in Murray Hill, New Jersey, poked and prodded the surface of a hunk of gray germanium—an element somewhat similar to its tin and silicon neighbors on the periodic table—with a battery power source to increase the output of an electrical charge. A few years later, in 1956, their work would earn them, along with team leader William Shockley, the Nobel Prize.

What Bardeen and Brattain had done was to "dope" or apply impurities to the germanium. So, depending on what impurities were added, the crystalline structure had either an excess of electrons (called N-type for negative) or few electrons (P-type). If there was a weak current flowing through the circuit of the doped surface with an excess of electrons, you could enhance it by applying an additional charge. Conversely, you could block the current until current was applied by adding another type of impurity. So by stacking the doped surfaces in either a P-N-P or N-P-N configuration, the little devices could be turned into amplifiers or on/off switches.

Naturally, the transistor's development was not a pure science quest. The transatlantic telephone line between North America and Europe, known as TAT-1, required repeaters to boost the signal. It might have been possible to send a telegraph-quality electrical pulse around the world with a tiny thimble-sized battery, but telephone communications required repeaters that boosted the signal. The flexible repeaters designed by Bell Labs were eight feet long and spaced every thirty-seven miles or so when the initial two cables were put down in the mid-1950s. The British Post Office, responsible for telephone communications, had its own tube repeater. Still, both models relied on vacuum tubes—specially designed and reliable vacuum tubes, but vacuum tubes with a limited life expectancy that would eventually have to be replaced. Transistors held the answer.

Vacuum tubes, of course, could perform the same functions, but they required much higher voltages. The same amount of "work" could now be done with considerably fewer electrical power tubes required in a fraction of the space. And even better, the little sandwiches of semiconductive material didn't blow out like tubes. Batteries were back as a viable power source.

Several months later Bell scientists had a working device, and a memo was circulated to name the thing. Among the names in contention were "Iotatron," "Crystal Triode," and, of course, "Transistor." The term transistor was a combination of the words transconductance (or transfer) and varistor (a device used to protect circuits against excessive voltages).

In June of the following year, Bell Labs held a press conference at its offices on West Street in lower Manhattan. The press release read, in part, "An amazingly simple device, capable of performing efficiently nearly all of the functions of an ordinary vacuum tube, was demonstrated for the first time yesterday at Bell Telephone

Laboratories where it was invented." A demonstration was given, along with a lengthy technical description of the science involved.

At the time, not many outside the fields of science and technology realized the significance of what the Bell scientists had done. The *New York Times* didn't appear to have much enthusiasm for the new device, famously burying the announcement on page 46 in a regular column called "News of Radio." And even then, it wasn't the lead item, following an announcement that Eve Arden would be starring in a new show called *Our Miss Brooks*, " . . . playing the role of a school teacher who encounters a variety of adventures." Eve Arden had somehow upstaged one of the most important technological breakthroughs of the twentieth century.

Transistors fared only somewhat better in the now-defunct *Herald Tribune* and mainstream science and technology magazines like *Popular Science* and *Popular Electronics*. To be fair, the *New York Times* wasn't alone in its seeming indifference. Aside from the professional technical journals and a few hobbyist publications, which hailed the announcement with varying degrees of geeky enthusiasm, the overall response was one of muted, earnest, and perfunctory reporting. What the scientists at Bell unveiled at their press conference was not a product like stereophonic sound or CinemaScope images on the big screen that the general public could immediately appreciate, if not fully understand. This new electrical component was tiny and its applications seemed somewhat distant.

On the other hand, the military immediately understood the significance of the transistor and tried to get it classified as top secret. This was more than just institutional paranoia. The Cold War was beginning to take shape—Churchill had delivered his Iron Curtain speech in 1946 at Westminster College in Fulton, Missouri, and George Kennan, the American ambassador to the Soviet Union, transmitted his historic "long telegram" that would form the basis of a decades-long policy of containment of Soviet ambitions, while a year later, President Harry Truman signed into law the National Security Act of 1947.

With tensions between East and West mounting, a device that could work at the heart of weapons and communications systems without many of the shortcomings of vacuum tubes was immensely valuable. Fortunately, Bell Labs eventually prevailed, and the patent for Three Electrode Circuit Element Utilizing Semiconductive Materials was duly filed in 1948, number 2,569, 347.

For years following their discovery, transistors moved forward with incremental improvements. The first commercially available transistor came on the market from Raytheon around 1950, but manufacturers didn't line up to place large orders, though they did find some use in a handful of obscure industrial applications along with a few do-it-yourself kits that challenged adventurous hobbyists to test their soldering skills in a new format called circuit boards.

At least part of the problem was the fact that there really wasn't a clearly defined consumer market for transistors. In a portable radio they could eliminate the large A battery, but most radios plugged into the wall. Then, in 1952, the Sonotone Corporation, a manufacturer of hearing aids, became the first company to offer a consumer product using transistors—albeit in hybrid combination with subminiature tubes. Interestingly, this was done under an agreement with AT&T that provided royalty-free licenses to manufacturers like Sonotone in observance of Alexander Bell's devotion to the deaf. In a bit of historical coincidence, Bardeen's wife, like Bell's, was hearing impaired.

Two years later, Bell Labs had built the first all-transistor computer—TRADIC (Transistor Digital Computer or Transistorized Airborne Digital Computer) for the U.S. Air Force using more than 700 transistors and diodes and 10,000 germanium crystal rectifiers. The entire unit fit into just a few square feet. This was a big step forward in the emerging computer field. The state-of-the-art ENIAC (Electronic Numerical Integrator and Computer) was a monster nicknamed "The Giant Brain." Secretly commissioned by the military during World War II to calculate artillery tables, ENIAC needed some 18,000 vacuum tubes, 1,800 square feet, and constant attention by a dedicated staff to change the tubes, which blew out with maddening regularity.

Extending the comparison, the first microprocessor made by Intel, the 4004, introduced in the early 1970s, packed the equivalent of 2,300 transistors onto a single chip, while today's processors contain the equivalent of nearly 300 million transistors.

Part of the problem with early transistors was the price. Vacuum tubes were plentiful and inexpensive thanks to economies of scale that pumped millions of them out into the market every year. Tubes were available for under a dollar, while Raytheon's early transistor, for instance, sold for $18.00 a pop (a little more than $150.00 in constant dollars). Even in the boom years of the 1950s manufacturers were still keenly aware that most consumers were "price sensitive" when it came to household items. Another problem was quality control. Transistors were more difficult to manufacture with a much higher rejection rate than tubes coming off the assembly line, which only added to the cost of the "good" transistors.

What turned things around was not a public clamoring for products packed with transistors, but the military. By the early 1950s, the Pentagon began spending tens of millions of dollars on transistors, actually paying for the construction of a Western Electric plant in Pennsylvania and financing transistor production facilities at existing plants for General Electric, RCA, and Raytheon as well as Sylvania.

It was in the new and ambitious weaponry systems, some of them on the drawing boards since the 1940s, like the first surface-to-air missiles, called the Nike Ajax, that transistors found viable applications. This kind of government investment, similar to the $30,000 Morse managed to squeeze out of Congress for his experimental telegraph line, would be repeated again and again through the years in programs leading to the development of the Internet, GPS, and a host of other technologies.

DESPITE THE DRAWBACKS OF THE new technology, transistors dramatically expanded the kinds of work batteries could perform. Not only did they require less power than tubes, but a lot of them could be packed tightly together on a circuit board to create relatively small, complex devices. This was also a huge step forward in terms of manufacturing. The boards were originally assembled much like tube-based units by assembly lines, primarily made up of women, soldering the components into place. Then, in 1949, two members of the U.S. Army Signal Corps, Moe Abramson and Stanislaus F. Danko, developed what would become known as the "Auto-Sembly" process. To form the circuits, the transistor's wirelike leads were inserted into tiny precut holes in a circuit board, the ends clipped, and the board run over a bath of molten solder to make the contacts between components. Because of the way each board's composition was designed, the solder would stick and harden to form the circuits between the transistors' contacts on the underside, but would not adhere where it wasn't needed. Companies could eliminate long assembly lines of workers soldering individual connections.

Taken together, these advances made for more durable, compact electrical components, perfect for military gadgets intended to be shot from very large guns, carried by troops, or installed in planes or ships.

BATTERY TECHNOLOGY WAS ALSO ADVANCING. Even toys were becoming more sophisticated, and the old zinc carbon formulations were falling behind. When Lewis Frederick Urry was transferred from Eveready's Canadian to its Parma, Ohio, facility, his first assignment was to find a way to extend the life of the company's line of batteries. It was, he knew, a hopeless task. Faraday's Law again!

What Urry did was take up the cause of the alkaline battery, which had been around for years, but never as a consumer item. Edison himself had developed one for cars. But they remained far too expensive for everyday use. Running through the materials, Urry finally found a formulation that included a combination that worked. Where he succeeded was in making the switch to powdered zinc, rather than a solid piece of the metal.

The powder, he realized, offered more surface area for the chemicals to react. It was an innovative solution, but also very conventional when it came to power sources. Since Volta, scientists had been increasing surface area, first by adding disks to voltaic piles, then plates to trough batteries. The Smee battery featured a roughed-up surface and, in a manner of speaking, so did Planté's lead storage battery.

The first modern alkaline was born, with an estimated life span of forty times that of the zinc-carbon formulation. However, selling the idea to his superiors would prove more difficult. By way of demonstration, Urry used a standard D cell and his new formulation in a pair of toy cars that he raced around the company cafeteria. "Our car went several lengths of this long cafeteria," he said in one interview, "but the other car barely moved. Everybody was coming out of their labs to watch. They were all oohing and aahing and cheering."

# 16

## *See It! Hear It! Get It!*

> *"Wherever ya are, and whatever ya doin' I want ya to lay ya hands on the radio, lay back with me and squeeze my knobs. We gonna feel it tonight. This is the Wolfman down here with the donkeys."*
>
> *—Robert Weston Smith aka Wolfman Jack*
> *XERF deejay*

The first transistor radio available to consumers arrived on the market in October 1954, just in time for Christmas. Produced and marketed by an Indianapolis company called Industrial Development Engineering Associates (I.D.E.A.), the Regency TR-1 measured a compact 3 inches by 5 inches by 1.25 inches and weighed only 12 ounces. Although not particularly impressive by today's standards, when compared to the subminiature tube radios at the time, like Automatic Radio Corporation's Tom Thumb or Motorola's Pixie, which weighed in at a pound or more when loaded with batteries, it was downright tiny.

The Regency contained four Texas Instruments (TI) germanium crystal transistors and drew its power from a whopping 22.5-volt Eveready specialty battery used in hearing aids. It wasn't particularly energy efficient by today's standards, but it offered a choice of four colors—black, red, gray, and white, and unlike the Belmont Boulevard, featured a small speaker. An optional leather carrying case was available at $4.95 as well as an earphone for $7.95.

In the interest of historical honesty, it must be said that the little Regency TR-1 was not the first miniature transistor radio. There had been some experiments with the concept. For instance, in 1953, the Search and Intercept Department of the Signal Corps Engineering labs came up with a transistor radio that weighed in at just 2 ounces and measured 2 inches by $1\frac{1}{8}$ inches by $\frac{3}{4}$ inch in a clear Plexiglas case, a little larger than a Zippo lighter. More or less assembled completely by hand as a "one off" over a single weekend, the experimental radio was built as a flashy, though classified, demonstration of miniaturization. It offered an earpiece, but no speaker, and included a long trailing antenna. The military engineers built it to be worn on the wrist and called it the Dick Tracy, after the comic book character. According to legend, Chester Gould, the superdetective's creator, got the idea of a miniaturized wrist radio after visiting the lab of Al Gross, the inventor of the Joan-Eleanor system.

THE TRANSISTOR RADIO'S ENTRANCE INTO the consumer market was accompanied by a surprisingly low-key amount of promotion. Even by the standards of the day, the ads seem painfully bland. However, despite the stodgy advertising in publications like *Holiday* magazine and the tepid announcements in the daily newspapers, there was a genuine urgency attached to the little radio. TI had a lot riding on the small device, which had eaten up a good portion of the company's working capital.

Founded in 1930 as Geophysical Services Inc., the company had switched from supplying oil industry technology to a military contractor during World War II and was now aggressively positioning itself to enter a new field. TI's management had seen the future and was now promoting it with twelve ounces of brightly colored plastic that could fit in your shirt pocket. The radio itself was a not so subtle effort by the Texas-based company to advance the use of its own line of transistors for the consumer products of other manufacturers and position itself as the premier supplier in what was predicted to become an increasingly crowded field.

One of the major obstacles the I.D.E.A. and TI engineering teams faced was getting the price down on the four transistors used in the Regency. Eventually they settled on about $2.50 apiece (roughly $20.00 in constant dollars), with the wholesale cost of the transistors equaling nearly a quarter of the retail price. Manufacturing problems with processes and quality control were still a problem, though by 1954 difficulties with price and quality were well on their way to being solved.

The obscure midwestern firm was not TI's first choice of a manufacturer. It was only after the big players, like RCA, politely declined that I.D.E.A. won the contract. The company did have some experience in the consumer marketplace with its line of signal boosters for television sets that allowed for reception outside of normal broadcasting range—an issue for early television owners in rural communities—but lacked the distribution muscle or credibility of a name brand manufacturer.

The plan was to get the radio to market as quickly as possible, within a year. For the TI and I.D.E.A. team that meant designing a product that had never been built with what amounted to state-of-the-art components. Rather than use the kind of metallic chassis tube sets relied on and soldering all the connections by hand, they adopted a version of the printed circuit board similar to those used in proximity fuses. Other components had to be modified from existing stock or be custom made. In the end, the team managed to get a prototype together by the fall of 1954 with manufacturing rapidly following in time for Christmas.

Now TI was dependent on the little company from Indianapolis and its modest promotional efforts. Stores carrying the new radio received countertop display cases while distributors were issued working TR-1's with clear plastic backs to show off the miniature circuitry. Promotional display cards, the same size as the radio, were printed up, to remind consumers the TR-1 could fit in a shirt pocket. In November, the *New York Times* ran a small four-paragraph story under the headline "Tubeless Radios Due."

"See it! Hear it! Get it!" the newspaper and magazine ads commanded, but at $49.95 (more than $350 in constant dollars) the TR-1 sold fewer than 20,000 units during that first Christmas season. Whether it was the hefty price tag, the somewhat demure marketing campaign, or the perception that the diminutive radio was nothing more than a pricey version of inexpensive novelties like "The World's Smallest Radio!" crystal sets advertised in the back of comic books, the TR-1 was not an overwhelming success during its first few months on store shelves in the few key markets where it was sold. Even *Consumer Reports* was not thrilled:

> The Regency is not anywhere near as small in size as Dick Tracy's wrist radio. It will, however, fit in the pocket of a man's sport shirt or jacket. The $49.95 selling price puts this receiver in the luxury class, since relatively few persons will be willing to expend that amount for qualities—other than novelty and size available in many other portables at a far smaller cost . . . A vacuum tube radio would give better tone quality, less background noise, and somewhat easier tuning at a considerably lower price but would be at a disadvantage in size, weight . . . and battery consumption, compared with the Regency (which would warrant an A rating where these factors are paramount).

However, radio enthusiasts of the variety who assembled Heathkit systems and haunted the aisles of RadioShack embraced the TR-1, eager to own the first transistorized radio. The movie producer Michael Todd handed dozens of them out to cast and crew of his film *Around the World in Eighty Days.* And IBM's Thomas Watson bought a hundred of the small radios as Christmas gifts for his managers as a not particularly subtle reminder they had yet to come out with a transistor-based computer. It isn't difficult to imagine what was going through Watson's mind. IBM's 1955 Model 650 weighed almost three tons and required more than 2,000 vacuum tubes. It was very much the 1955 Cadillac of computers, tail fins and all.

The vast majority of American consumers, however, didn't know exactly what a transistor was supposed to do or why you would want to put a radio in your shirt pocket. In the minds of a good many consumers, batteries were still power for the simple or inexpensive. They powered toys for children, flashlights for emergencies, and not much else. Serious consumer items, televisions, record players, and family radios plugged into the wall and relied on tubes that came to life slowly with a soft and reassuring amber glow amid complicated wiring bustling around a galvanized steel chassis. The fact that the designers had chosen brightly colored plastic didn't help matters. Serious electronics—real state-of-the-art stuff—came in a somber black and brown. A good portion of quality home electronics were still housed in wood cabinetry and resembled furniture. And why would you want something small that consumed less power when there were so many electrical outlets in the world to plug into?

Compared to today's consumers who find added value in compact cell phones and notebook computers, the typical 1950s consumer saw the perceived value of a device actually decrease along with its size in an age with no clear concept of personal electronics. Bigger was pretty much always better.

The TR-1 did eventually find its customer base, and it wasn't with the readers of *Holiday* magazine or even those who cared how transistors actually worked. Teenagers would prove to be the early adopters of the new technology. Popularity of the little TR-1 grew among young people so that by the end of 1955, I.D.E.A. had sold more than 100,000 units. Teens in New York City were soon listening to Chuck Berry's "Maybelline," The Penguins's "Earth Angel," and Fats Domino's "Ain't That a Shame" on WINS, one of America's first rock 'n' roll radio stations. In Southern California, the colorful plastic radios became a fixture on beaches and at poolsides.

The popularity of the TR-1 spread quickly, and the radio became a popular birthday and high school graduation gift. Soon farm kids were hanging the little radio on their tractors and listening to weather reports or country western hits like Tennessee Ernie Ford's "Sixteen Tons." Eventually, the transistor radio itself would work its way into songs such as Buck Owens's hit "Made in Japan" (1972), The Beach Boys's "Magic Transistor Radio" (1973), Connie Smith's "Tiny Blue Transistor Radio" (1965), and, of course, Van Morrison's "Brown Eyed Girl" (1967).

THE SINGLE EARPLUG THAT ALLOWED for private listening—a relatively new concept in the 1950s—became a standard comedic device among cartoonists and situation comedy writers. And, too, there was something sneaky, if not outright subversive, about listening to music or sports broadcasts privately with the earphone while out in public. It was frowned upon in much the same way that texting in social situations is seen as ill-mannered.

Nevertheless, the transistor radio found its home in the emerging youth market of the 1950s and 1960s as the postwar baby boom was just getting under way. Competition entered the market, and their ads didn't feature middle-aged bow-tied business executives and women in evening gowns. These new ads offered up wholesome images of teens in pressed pants and sweater sets or vaguely "beatnik" youths in rumpled chinos, turtlenecks, and Ray-Bans.

Popular culture was coalescing around the baby boomers. Rock 'n' roll was a defining part of that youth culture and, powered by little 9-volt batteries, which soon became known as "transistor radio batteries," pocket radios allowed teens to carry *their* music with them. The transistor radio, like the car, represented a form of independence that cut teens free from the domesticity of the living room radio and connected them to a larger world. A few years later, at night, when the radio waves bounced off the hardened ionosphere, the little radios in New York City, Des Moines, Iowa, and Klamath Falls, Oregon, could pick up outlaw radio stations just across the Mexican border, beyond the reach of the FCC. Broadcasting at a continent-blanketing, bird-killing 250,000 watts, Robert Weston Smith, known to his loyal XERF listeners as Wolfman Jack, gave countless teens—far from the major radio markets—their first taste of Howlin' Wolf, James Brown, and rock 'n' roll on their little transistor radios.

This was not *American Bandstand, The Ed Sullivan Show,* or suitable for the family radio in the living room. Listeners didn't even

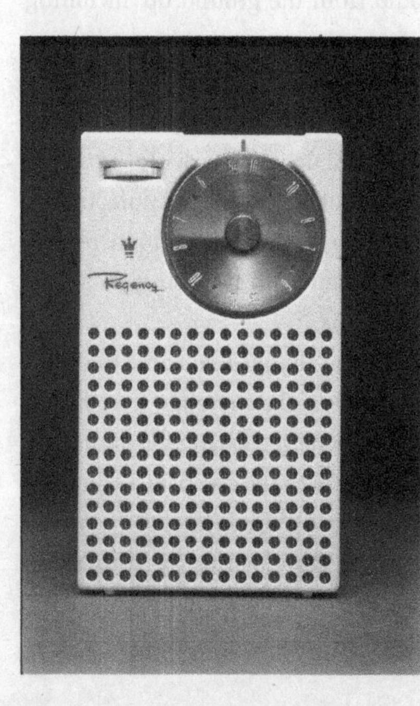


*The Regency pocket radio, produced by the small Indianapolis-based company I.D.E.A. and Texas Instruments, was not a resounding success, but proved that transistors had a place in consumer products.*

know if Wolfman Jack was black or white. Naturally parents hated him and so did the Soviets, who took to jamming his decadent American broadcasts that originated from Ciudad Acuña, Mexico (then called Villa Acuña) just across the border from Del Rio, Texas. But he was perfect for the transistor in the teenager's bedroom with its little earpiece.

Suddenly the battery was back on the leading edge, at least in the eyes of consumers.

NOT LONG AFTER THE INTRODUCTION of the TR-1, a Japanese company barely ten years old and armed with a license to manufacture transistors from Western Electric began building its own line of transistorized radios. Tokyo Tsushin Kogyo (Tokyo Telecommunications Engineering Company) began with the TR-55. Much larger than the TR-1, the TR-55 sold for about $30.00, but was only available in Japan. As a point of pride, Tsushin Kogyo was the first company to manufacture the radio from the ground up, including transistors. Not a bad start for a young company that was still earning its reputation building tape recorders.

For its next model, the TR-63, designed as a "pocketable radio," the small company gambled and turned down an order for 100,000 units from Bulova that would have put the watch manufacturer's own "brand name" on the unit. It was, by any standard, an incredibly risky proposition for a Japanese company with no history in a U.S. market filled with World War II veterans.

For these first exports, the company changed its name to something Americans could easily pronounce and remember. Combining the Latin word for sound (*sonus*) with American slang for young boy (sonny), the Sony Corporation was born with the company's cofounders Masaru Ibuka and Akio Morita rejecting an addition to the name, such as "electronics" or "radio," on the grounds that the future was still uncertain. Needless to say, the gamble paid off handsomely.

It was Sony's TR-63, introduced in 1957 and selling for $39.95

(about the average month's salary for a Japanese worker) that really blew the lid off transistor radios. Strictly speaking, Sony's entrant into the field was not actually "pocketable," since the final product was just a little bit too large to fit into the standard-sized shirt pocket by a few centimeters. Morita, who would later invent the Walkman, quickly solved the design problem by issuing shirts with slightly larger pockets to his sales force.

There were a few other problems solved along the way as well. The TR-63 used six transistors—compared to the four used in the Regency—for better reception while consuming only half the power and used a small, 9-volt battery that became standard for transistor radios. Very much a "second generation" product, the TR-63 set the standard for transistor radios for years to come.

It almost goes without saying that no technical advance comes to the public's attention without attracting its share of crackpots and crackpot theories. So it was (and actually still is) with the transistor. The most entertaining of these theories is that the transistor is actually not a product of Bell Labs, but a successful top-top-top secret effort in reverse engineering of alien technology salvaged from a crashed spaceship. And while scientists and engineers truly hate the theory, it's still noteworthy for a number of very good reasons.

The alien concept echoes the myths of centuries past when tales of miraculous technologies and strange natural phenomena spread throughout Europe. In the new myths, foreign lands have become distant planets. And, by refuting the myth of what amounts to extraterrestrial patent infringement, it forces us to examine the way the transistor came into the world. Looking back, there's no mystery: the transistor was developed in a series of incremental steps dating back to the very early discoveries of Karl Ferdinand Braun.

During World War II, the Rad Lab at MIT and Purdue University were frantically working to improve the reception on radar systems and began looking at different materials, including germanium, which was added to the list of semiconductors consisting of silicon, selenium, and tellurium in 1926. Unlike others on the short list, germanium could be refined down to a very pure state. This initial wartime effort by the nearly forgotten Purdue team, made up of graduate students and led by Dr. Karl Lark-Horowitz, pioneered the ability to produce very pure germanium and provided a better understanding of the material. After the war, Purdue researchers continued their semiconductor explorations, though in a more academic setting than Bell Labs. By some accounts, the Bell Labs team beat them to the invention of the transistor by just weeks.

Nearly parallel work was taking place in Germany during World War II, specifically by two physicists, Herbert F. Mataré and Heinrich Welker. Apparently radar was not a priority for the Germans, at least not at the beginning of the war. "My pilots do not need cinema on board," the Luftwaffe's Hermann Göring was reputed to have said, no doubt flush with early ground victories and fondly remembering his own days as a dashing pilot flying biplanes during World War I or working in air transport between the wars. Fortunately, Göring's cavalier dismissal of the new technology would prove disastrous as cloud cover and poor weather conditions became meteorological allies for increasingly aggressive British bombing raids inside Germany. By the time it became apparent the Nazis needed the technology, it was too late to play catch-up.

However, for scientists like Mataré and Welker, the futile push toward an advanced radar system led them down the same research road as the Rad Lab, providing valuable insight into semiconductors. Following the war, the pair went to work for Westinghouse in France where they continued research on the transistor. Eventually, they independently developed a device almost identical to what the team at Bell Labs created, just a few weeks after Bell's scientists. They called the new device a "transistron." It was only months later that news of the American version reached them.

Mataré, in particular, seemed destined for bad timing. Returning to Germany in the early 1950s, he formed his own company, called

Intermetall. Looking to put the transistor to work, he produced a prototype transistor radio and launched the concept at the Düsseldorf Radio Fair in 1953, a full year before the Texas Instruments/I.D.E.A. hit the market. "It was received very well," Mataré said in one interview. "People were amazed by its size." And then the company backing Intermetall cut its funding, and the little radio never progressed beyond prototype stage.

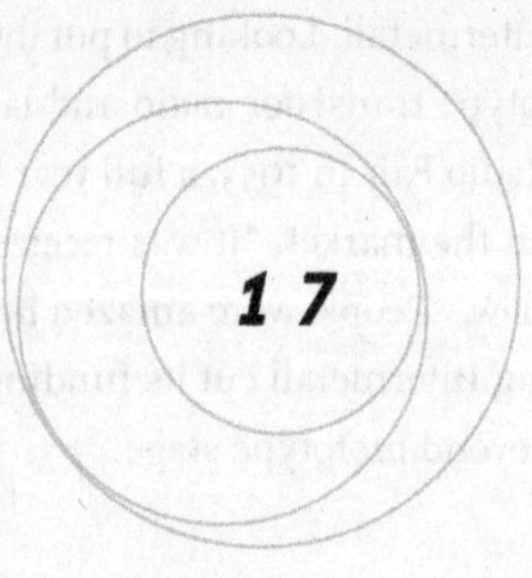

# Smaller and Smaller

*"The future of integrated electronics is the future of electronics itself . . ."*

—*Gordon Moore,* Electronics *magazine, 1965*

The reign of the vacuum tubes lasted nearly half a century until made obsolete by transistors. Although small, reliable, and energy efficient—at least compared to power-gobbling tubes—transistors remained state-of-the-art for less than a decade. The rapid growth in the complexity of electronic devices challenged transistors' practicality in much the same way vacuum tubes had been stretched to their limits. Computers, for instance, sometimes requiring upward of 100,000 diodes and 25,000 transistors, were becoming hugely expensive to manufacture. IBM's 7030 ("Stretch")—the company's all-transistor supercomputer conceived in the mid-1950s—used an astonishing 170,000 transistors wired into circuit boards.

And the military, at an early critical juncture in the Cold War, needed new technology for advanced weapons systems. In 1951, the navy briefly sponsored something called Project Tinkertoy, dreamed up by Robert Henry at the National Bureau of Standards. The idea was as simple as it was clever, and the navy thought it had real potential. Small ceramic wafers that snapped together, each about seven-eighths (some documents list it as five-eighths) of an inch square, would house a different type of standardized component. The idea was to cut production time and cost by using automation to attach the transistorized components to the wafers, which could then be quickly assembled into complete units. For whatever reason, the Tinkertoy concept faded quickly, but not before the navy invested nearly $5 million in the effort.

The U.S. Army Signal Corps followed a short time later by investing heavily in a more sophisticated concept called Micro-Modules proposed by RCA. Like Project Tinkertoy, the Micro-Module effort also included ceramic wafers, though slightly smaller versions measuring a little more than a third of an inch square and a hundredth of an inch thick and holding multiple components. As with Tinkertoy, the modules were essentially tiny circuit boards, but they pushed circuit density to impressive new levels. As a demonstration, RCA built a radio into a fountain pen—the ultimate transistor radio—and the army brass loved it. Clearly this was the future for transistors.

But the Micro-Module project, though very clever, was already obsolete by the time it was announced in 1959. In fact, the timing could not have been worse. RCA and the Signal Corps' joint announcement at the Institute of Radio Engineers (a predecessor component of the Institute for Electrical and Electronics Engineers, or IEEE) convention was overshadowed by Texas Instruments's debut of the first integrated circuit (IC)—the computer chip. And the computer chip was not an incremental advance, but a substantial technological leap forward. Not only was it significantly smaller than the transistor, but almost from the start its potential seemed nearly unlimited.

TI wasn't the only player in the new field. Fairchild Semiconductor—financed through the defense contractor, Fairchild Camera and Instrument Corp.—was also in the game. In fact, Fairchild's patent for the technology was filed *before* TI's by several months, but it was held up because of wording. TI had simply written its application more narrowly, speeding the approval process. Of course, the lawyers became involved and the case dragged on for a decade, eventually ending up in the U.S. Court of Customs and Patent Appeals (which ceased to exist in the early 1980s). After much legal wrangling, the court upheld Fairchild's patent claim while simultaneously giving TI the credit for building the first integrated circuit. But in the end, of course, it didn't matter. Integrated circuits had arrived and were clearly the future.

As with the first transistor, the military found immediate use for the technology even as chips from both companies began rolling off the line in 1961. And if anyone doubted the potential of the new technology, those reservations were soon put to rest when TI unveiled what it called a molecular electronic computer. Built under contract for the air force in 1961, the diminutive computer measured just 6.3 cubic inches and weighed in at 10 ounces. The unit didn't include any kind of user interface—neither a screen nor a keyboard—but it got the point across. As TI proudly noted, the miniature package included 47 chips that did the work of about 8,500 transistors, diodes, resistors, and capacitors.

A year or two later, ICs were built for the Minuteman missile systems of the early 1960s as well as for NASA, which was responding to President Kennedy's challenge to put a man on the moon by the end of the decade.

VERY SOON BATTERIES WOULD BE powering very small components doing highly complex calculations. Just as the telegraph had been responsible for the death of distance by severing the connection between the flow of information and travel time, the IC shattered the long-standing relationship between the complexity of a task and the size of the device performing it. Small and portable devices could now be built to perform extremely complex work quickly. This was made clear a few years ago when a group of electrical engineering students at the University of Pennsylvania decided to commemorate the fiftieth anniversary of ENIAC by replicating its entire processing architecture, all 30 tons of 18,000 vacuum tubes, 7,200 diodes and 1,500 relays on a single computer chip that measured a petite 7.44 by 5.29 square millimeters.

WRITING FOR *ELECTRONICS* MAGAZINE IN April of 1965, Gordon Moore, who was still heading Fairchild Semiconductor's R&D effort, before leaving to cofound Intel, predicted a doubling of circuits in ICs every two years and saw no reason why that shouldn't continue far into the future. "The complexity for minimum component costs has increased at a rate of roughly a factor of two per year," he wrote. "Certainly over the short term this rate can be expected to continue, if not to increase. Over the longer term, the rate of increase is a bit more uncertain, although there is no reason to believe it will not remain nearly constant for at least 10 years. That means by 1975, the number of components per integrated circuit for minimum cost will be 65,000."

The landmark technical essay formed the basis of what's become known as Moore's Law—an informal prediction that the number of processing components on a chip doubles every two years. In a very real sense, *all technology* is interim technology, but some is clearly more temporary than others. What Moore predicted correctly was that ICs would not become the 8-track players of computing.

"The future of integrated electronics is the future of electronics itself," Moore wrote in the 1965 article.

> The advantages of integration will bring about a proliferation of electronics, pushing science into many new areas . . . integrated

> circuits will lead to such wonders as home computers—or at least terminals connected to a central computer—automatic controls for automobiles, and personal portable communications equipment. The electronic wristwatch needs only a display to be feasible today.

Yes, there were electronic wristwatches, but what Moore was referring to was the problem of a practical, small-scale user interface capable of matching the IC's data output. For large computers, you could always use a CRT, like a television screen or printer of the kind employed by Western Union at the time, both unrealistic solutions for a portable device. And, since the concept of portable implies that the power supply is a convenient size as well, something small and less power thirsty would be needed. The answer, interestingly, arrived nearly simultaneously with the publication of Moore's essay. Researchers at RCA's lab in New Jersey were on the verge of a major breakthrough in liquid crystal displays or LCDs.

The LCD can actually be traced to the late nineteenth century, when the Austrian botanist Friedrich Reinitzer happened to notice that some organic crystals—cholesteryl benzoate—exhibited strange properties when exposed to heat. They turned cloudy and then clear at specific temperatures. That is to say, they acted unexpectedly, but consistently, which is always of interest to scientists. He related the discovery to a professor of physics, Otto Lehman, who carried on the research just long enough to note some interesting refraction properties and came up with the name *fliessende Kristalle* (liquid crystal). The substance was a scientific curiosity, but not much more, until the scientists at RCA got hold of it in the early 1960s.

As it turned out, the liquid crystals reacted not only to heat, but also to an electromagnetic field. So, if you squished the crystals between two panes of thin glass with a conductive surface and applied a relatively small amount of power, they would align and become opaque. Treated with the right kind of dye, they even changed color in a predictable manner.

A few years later, RCA showed off a very primitive version of an LCD and a window that went dark when current was applied. And then . . . nothing. The powers that be at RCA were far from enthusiastic. "The people who were asked to commercialize [the technology] saw it as a distraction from their main electronic focus," said George Heilmeier, one of the LCD proponents who left RCA when the project languished and eventually went on to head Defense Advanced Research Projects Agency (DARPA), the military's primary research and development organization and the agency responsible for the early development of the Internet.

At the time, RCA was one of the most successful companies in the world with a good portion of the cathode ray tube (CRT) market and little interest in pursuing something so esoteric. The whole enterprise lay dormant until 1968 when a Japanese television crew filmed Heilmeier demonstrating his proof of concept LCD for a documentary called *Firms of the World: Modern Alchemy.*

A year later an engineer at Sharp Corporation recognized the technology as a possible solution to a pocketable calculator display on the drawing boards. RCA, apparently still not interested in the technology, offered little assistance. So, with the American company proving uncooperative and virtually no current material published on the obscure field, the engineers at Sharp did what any good engineer would do—they watched the videotape of Heilmeier's demonstration. Although all the bottles at the lab had their labels conveniently turned from the camera, the engineers were able to gather enough clues to at least begin their research and somehow managed to perfect the technology within a relatively short amount of time.

Then, in May of 1973 Sharp introduced the Elsi Mate EL-805 pocket calculator to the world. The unit, which housed five ICs, was less than an inch thick and weighed just 7.5 ounces. But the real surprise came in the fact it could run for a hundred hours on a single AA battery. That is to say, power consumption was estimated at

1/9000 of other battery-powered calculators on the market. LCDs had solved the problem of the power-hungry user interface.

By any standard, the Elsi Mate was a breakthrough, far surpassing even TI's collaboration with Canon on a "portable" calculator introduced in 1970 called the Canon Pocketronic, which weighed in at nearly two pounds and was anything but pocketable at a bulky 4 by 8.2 by 1.9 inches. The TI unit didn't even have a screen, but rather a thermal paper printer. Users could read the paper printout behind a small glass magnifying window.

The Datamath or TI-2500, introduced in 1972, was better, measuring 3 by 5.5 by 1.7 inches and offering an LED display, but it needed a half dozen AA batteries to power up. The Datamath was followed in the calculator market by the Japanese firm Busicom with its LE-120A or "handy," which needed only four AA batteries to power its LED display.

And there was a problem. LEDs, though far more energy efficient than incandescent bulbs, were still relatively piggish about power consumption when it came to small battery-operated devices. This was made painfully clear when Hamilton announced its Pulsar digital watch—the first watch with no moving parts— to great fanfare in 1970. Called a "time computer," the Pulsar was proudly promoted as state-of-the-art. With the equivalent of 1,500 transistors in its ICs, all dedicated to keeping perfect time for the owner, it was well worth the estimated $2,100 price tag (about $10,000 in constant dollars).

Except that even before the first watch was shipped to stores, engineers discovered that the LEDs were draining the batteries at an astonishing rate. According to Hamilton at the time, it was the first use of LEDs in a wristwatch and mistakes were natural. The ICs worked fine, keeping more or less precise time, but it took two years for the company to solve the user interface problem with the LEDs.

First, the engineers replaced the originally planned single silver zinc rechargeable battery with two replaceable power cells (a certificate for a free second set was included with the purchase). To save power the user had to push a button redesigned to light the LEDs on the face. According to the manual, you could push the button twenty-five times a day and the watch would run for a year. It wasn't the most elegant engineering solution, but the novelty of the watch seemed to outweigh the small inconvenience. Johnny Carson proudly displayed one on *The Tonight Show,* Richard Nixon was said to have worn one, and James Bond wore one, albeit briefly, in *Live and Let Die* (1973).

A year later, Seiko came along with a power-sipping LCD watch that set the standard with a continuous time readout. A nearly perfect match to the lower energy requirements of ICs, within a decade LCDs began appearing everywhere, becoming the user interface of choice for a new generation of portable gadgets.

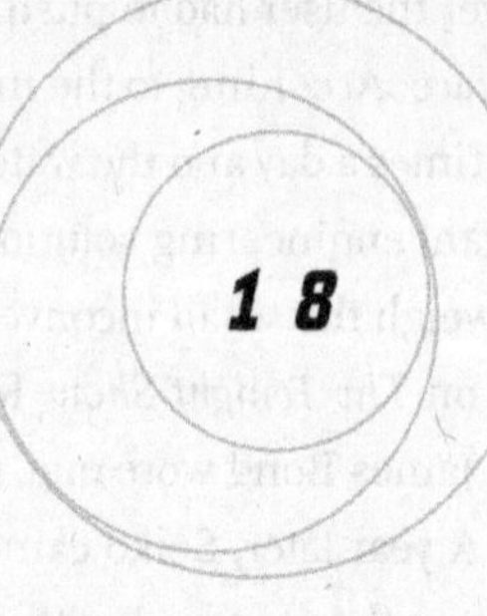


# Always On

*"Lashing out the action, returning the reaction . . .*
*Battery is here to stay!"*

*—Metallica*

Through the 1980s and 1990s batteries seemed to keep pace with all of the consumer gadgets and gizmos entering the market. New battery chemistries were coming into use, and the technique by which manufacturers rolled battery components tightly to increase density and create the different cells—known in the industry as a jelly roll—allowed them to pack more and more power into smaller packages. A jelly roll D cell can increase surface area to an impressive thirty inches. For a long while, this seemed to work and batteries settled into five types of rechargeable chemistries for consumer electronics: the alkaline, nickel cadmium (NiCd), nickel-metal hydride (NiMH), and lithium-ion (Li-ion).

Electronic devices evolved at an amazing rate during the 1980s and 1990s, driven, at least in part, by Moore's Law. Battery engineers, still working under the iron rule of Faraday's Law, weren't as lucky. They found themselves constantly trading off between energy density, longevity, and size. The problem was keeping up as the use of portable products increased.

Alkalines are more efficient than standard dry cells, but are primarily suited to low-drain tasks like television remotes and toys. With devices such as portable music players seeing increased use, what was needed was a rechargeable battery.

The first generation of serious rechargeables included the NiCds, which were popular for a while; however, they were not only toxic but also prone to the notorious memory effect—if you charged them before they were completely discharged, their ability to hold a charge would diminish precipitously. Fortunately for all concerned, they're in the process of being phased out in favor of the more environmentally friendly NiMH for things like digital cameras and electric razors.

AND THEN THERE'S LITHIUM, WHICH now powers a good portion of our electronics but has always proved problematic. First discovered in 1800 in a Swedish iron mine in the form of petalite ore or lithium aluminum silicate, lithium came to the attention of scientists in a less than pure state. It took more than a decade before a young chemist, Johan Arfwedson (sometimes Arfvedson), working in the lab of Jöns Jakob Berzelius, famous for figuring out atomic weights and devising the system of chemical notation we use today, to even classify it. Because the substance seemed alkaline, he gave it the misleading name *lithos* (Greek for stone) to distinguish it from salts found in organic matter, like plants. In fact, it was an alkali metal.

Not much happened until Humphry Davy with his supercharged voltaic pile, and another member of the Royal Institution, W. T. Brande, managed to isolate it through electrolysis. What they found when they applied a good jolt of electricity to lithium chloride was a very reactive silvery metal that was highly flammable and quickly oxidized when exposed to air. The lightest, least dense solid metal, lithium is very much a metal with attitude. It didn't take long to realize that you couldn't leave the stuff lying around the lab like, say, lead or copper. You had to store it in oil.

Chemists loved lithium, calling it the "mystery metal," though its uses seemed limited. It wasn't until the 1960s that the idea of the lithium battery gained some traction for use in pacemakers. Then, in the 1970s Exxon researchers began working on lithium batteries in earnest, as did Lew Urry, who already had the alkaline battery to his credit at Eveready.

The potential benefits of a lithium battery were obvious—high voltage, "high energy density," and a chemical side step around Faraday's Law that was holding traditional batteries back. The possibility for extended battery life and higher charges was there, but was it worth the effort? Did the consumer market really need a new battery, particularly one made up of this strange, highly flammable metal? After a few initial efforts, America's leading battery companies abandoned lithium altogether. Transistor radios, flashlights, and all manner of toys ran just fine with the standard batteries. Going lithium would require retooling plants or making enormous investments in new plants. And, too, lithium wasn't the kind of material you want stored in a warehouse. For battery manufacturers it was an easy decision: let someone else do it.

It was around that time that the government, the army, the navy, and even NASA stepped in by funding research while the FAA established safety guidelines. If high output and long life weren't absolute necessities for consumer products, they were certainly welcomed for things like emergency locators for planes and satellites and on the battlefield. Unfortunately, lithium had not gotten any easier to work with over the years. Several fires and at least one death were reported. Progress was made, but the batteries remained a highly specialized product. That is, until Sony and the Asahi Chemical Company entered the picture.

Picking up the research where American scientists left off, they brought the first lithium-ion (Li-ion) battery to market in the early 1990s. Very much the right battery at the right time, the Li-ion batteries presented a major technological advance. Not only did they not contain lithium in its dangerous form—just the ions—they also produced a solid-state chemical reaction, which meant very little self-discharge. Engineers began calling the new configuration a "rocking chair" battery for the way the lithium ions rocked back and forth between the two electrodes. They could sit around for a long time before going "bad." They seemed a perfect match for the new age of portability that very rapidly evolved from the AA-powered Walkman to laptops, cell phones, iPods, and PDAs. Lightweight and moldable to fit most devices, Li-ion batteries also don't suffer from the dreaded memory effect. Although they generally don't last beyond three years, neither do most of the products they power.

In a little more than a decade, extended life and higher charges had progressed from a nice luxury to a decisive issue. Sony was soon joined by South Korean manufacturers and even Chinese companies as the Far East became the center of Li-ion battery manufacturing.

What is interesting is that American companies were not "beaten" at the rechargeable battery market by low wages, but seem to have made the decision not to aggressively participate on a large scale. Some place blame on the vertical integration of the Asian electronics manufacturers or the low profit margins compared to primary batteries.

This state of affairs caused no little concern among electronics manufacturers in the United States. While America continues to act as an innovator of the new technologies, there is some question as to how long that will last. Asian manufacturers have not only perfected the manufacturing processes but are also funneling significant amounts of capital into their R&D programs. All of this has happened quietly, in large part because batteries are increasingly black-boxed, hidden away from the consumer, without brand names or the benefit of advertising. Still integral but unseen, batteries are on the verge of some very large technological advances.

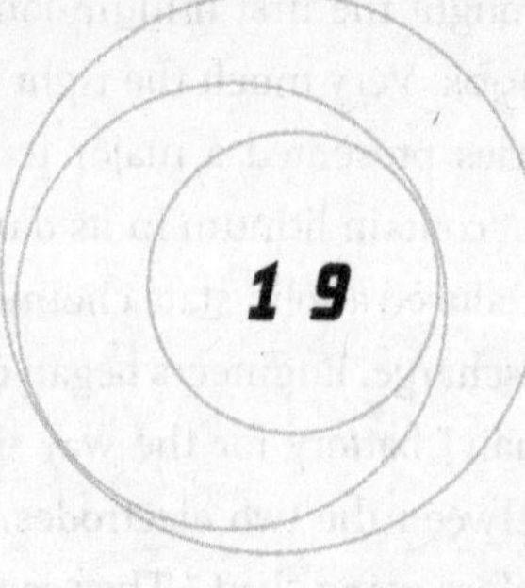


# Lab Reports

> *"The past is a foreign country: they do things differently there."*
>
> —*L.P. Hartley*, The Go-Between

What has changed over the past decade is that battery technology is now in the midst of an aggressive evolutionary process, pulled by increasingly sophisticated devices as well as by advances in pure science. For the consumer market, the goal is to build batteries that deliver greater amounts of power at higher levels and that recharge quickly. Even as consumer products have become more sophisticated, offering more energy-burning features, systems designers have struggled to keep pace, incorporating low-power modes into the IC circuitry that shut down specific functions when not in use, essentially updated technical versions of what

Hamilton's Pulsar engineers came up with by requiring a press of the button to read the LED time display.

The basic principles of battery chemistry, essentially unchanged for 200 years, have had little reason or opportunity to evolve. Consumer gadgets, by far the largest users of batteries, have more or less been designed with available and proven power sources in mind. What has changed over the past decade is the way consumers use their batteries in an ever-growing number of portable devices. In our current age of portability and increasing technological sophistication, power sources have become the weak link. Even the most advanced configurations available are very much early to mid-twentieth century chemistries powering twenty-first-century technology. The functionality and user interface of an iPod or a cell phone would probably amaze or baffle Alessandro Volta, though he would likely have little trouble grasping the basic design principles of its battery.

Of course, there have been recent success stories. The batteries used in both Gulf Wars, known as the BA 5590, has been powering up an ever-wider range of military electronics, like GPS, portable targeting systems, night vision, and portable computers for more than a decade. This is a long way from the simple flashlights and walkie-talkies of World War II. Weighing in at a little over two pounds, the BA 5590 comes in three energy flavors: Lithium Sulfur Dioxide, Lithium Manganese Dioxide, and Lithium Rechargeable (Li-ion).

So critical were the batteries, according to reliable reports, the second Gulf War was nearly halted in 2003 due to lack of batteries at the front line. It was, according to one U.S. military official, a "near-term disaster," with the military within days of depleting its supply. What saved the day was fast thinking that saw additional batteries airlifted into the theater of operations and a quick end to the ground fighting.

The military, which has taken battery technology seriously since the Civil War, has recognized the need for even more efficient batteries and may just produce the next generation of power sources that fire up consumers' MP3 players.

And then there's the Hubble Space Telescope. Launched in 1990 with six, 125-pound rechargeable nickel hydrogen batteries that add up to about 460 pounds and measure 36 inches long, 32 inches wide, and 11 inches high, scientists and engineers estimated their life expectancy at around five years. In fact, they lasted some nineteen years before showing signs of diminished charging capacity. NASA attributed their world-record-setting longevity to careful management during the charging cycles along with unusually tight engineering specs. The new batteries, which were installed in 2009, are also nickel hydrogen, but manufactured using an assortment of new processes. The compounds are poured into their molds in a sludgelike consistency, and then essentially baked to eliminate the water. According to NASA, the new batteries should remain fully functional until 2013.

*The six, 125-pound rechargeable nickel hydrogen batteries that powered the Hubble Space Telescope set a record for longevity. Solar charged, they were packed three to a module (one module is pictured above). Initially calculated to have a five-year life span, the batteries were nursed along by NASA engineers for nineteen years. They were eventually replaced in 2009 by a new set of nickel hydrogens that employed new manufacturing processes. According to NASA they should remain functional until 2013.*

What does seem apparent is that portable power sources are on the verge of some very large changes. Battery life in our increasingly portable world has become a competitive issue. A notebook computer that delivers twenty hours of battery life at a stretch would have a distinct advantage in a marketplace where even the most sophisticated products are seen as commodities.

ACCORDING TO SOME EXPERTS, SIGNIFICANT improvement in traditional battery technology may be coming to an end. The last major breakthrough, lith-ion batteries during the 1990s, say some experts, brought the industry close to the end of the line of usable materials. We are, they say, at the point of incremental improvements.

One of the more promising technologies, the methane fuel cell, has been aggressively pursued by Sony. The hybrid version unveiled featured a miniature fuel cell, small enough to fit on a keychain along with a Li-ion battery. The unit, Sony pointed out, can either switch between the battery and fuel cell or run both systems simultaneously to power small devices.

Flat film batteries, already on the market, offer another solution. Flexible and no thicker than a typical playing card, they use somewhat standard chemistries in a new way. They are made up of micron- and submicron-thin layers that create the anode, cathode, and electrolyte, and researchers have to date gotten them down to about five microns or 0.00019685 of an inch thick to produce an electrical charge. While not suitable for typical consumer products, they offer enough power to run a small IC in your credit card or label on canned peas, or even small active radio frequency identification (RFID) tags, store data, or power up some basic IC hardware. This technology is already offered by TI and a few other companies for specialty applications, such as Micro-Electro-Mechanical Systems, or MEMS, that require relatively little power. And when combined with a new generation of flexible ICs, due out soon, computing power will migrate from hard, protective coverings to a wider range of applications, such as clothing capable of powering monitors for heart rate or lighting for some form of decorative display.

THE WAY WE CHARGE BATTERIES is also due to change in the near future. Plugging in a device to recharge batteries is on the way out with a variety of new technologies on the horizon. Nikola Tesla's once fantastic-sounding dream of remote energy transmission is quickly becoming a reality. The most practical method works through a kind of induction coil that beams energy into a receiver mounted in the device, though this only works for relatively short distances or when the device is placed directly against the coil. Even more intriguing is the concept the engineers at Nokia are currently working on, which is a way to harvest ambient electromagnetic radiation emitted from things like Wi-Fi transmitters and cell phone towers that surround us every day, filling the air with energy to recharge batteries or run small devices.

# EPILOGUE

## *Bring on the Future*

*"Prediction is very hard, especially when it's about the future."*

*—Yogi Berra*

Predictions far into the future are always dangerous. This is painfully evidenced by the legions of baby boomers who have grown to middle age and beyond still waiting for the arrival of flying cars and personal jet packs promised in the magazines of their youth.

In the more distant future are ultracapacitors or double-layer capacitors or electric double-layer capacitors. A major technological leap forward from standard batteries, ultracapacitors work not by generating a charge through a chemical reaction but by holding an electrical charge in much the same way as a Leyden jar. In their most common form, ultracapacitors are made up of two nonreactive plates coated with carbon mounted in an electrolyte. The trick is that the surface area of each porous plate is enormous, so it's able to hold a large charge.

Technically, the Leyden jar was a capacitor, though ultracapacitors first arrived on the scene in the late 1950s with a somewhat uneven improvement rate over the years and are today used in a variety of applications, such as supplying backup power for electronic devices. However, what's changed is experiments in materials that supply an enormous amount of surface area, hence more electricity than standard capacitors, but still significantly less than a typical Li-ion battery. The surface area just isn't there yet, but they're getting close. In theory, there are huge advantages to this kind of technology, at least with a few more product generations. For one thing, because there isn't a chemical reaction, ultracapacitors are less susceptible to environmental factors. They can also deliver more power more quickly than batteries and, since they don't depend on a chemical reaction to recharge, they recharge quickly, within seconds. In theory, you could fully charge your iPod while waiting for an elevator or in line at the bank. They can also be recharged more often without losing their efficiency, say 50,000 times or more.

In the vanguard of the research effort are Professor Joel E. Schindall and a team at MIT's Laboratory for Electromagnetic and Electronic Systems (LEES). What they've done is replace the activated carbon of ultracapacitors with carbon nanotubes that are 1/30,000 of a human hair. The tubes are actually "grown" on the surface. Once a plate's surface is prepared with a catalyst, it's then exposed to a hydrocarbon gas at a high temperature. As the gas fills the closed chamber, the catalyst captures the carbon atoms to build up or self-assemble a fuzzy layer of tiny regularly spaced nanotubes vertically aligned on the plate's surface within minutes to grow a little nanotube forest. "What we're trying to do is grow an array of nanotubes on a conducting electrode, like tin foil, for example," Schindall explained. "We believe a nanotube could operate at three to four volts, about five times as much storage capacity as a commercial ultracap."

This is the kind of technology that could not only find use in small, portable devices but also in autos and storage for alternate energy generators, such as wind power. If all goes well, commercial products could be ready within a decade. However, more than a decade has passed since the introduction of Toyota's Prius, the first mass-produced hybrid car, with only incremental improvements to the original concept. Compared to the technological enhancements the iPod has undergone since its introduction a few years after the Prius, the electric car appears to be moving at a snail's pace.

The comparison, of course, is not a fair one. However, it is those very elements that make it unfair—all those differences between the car apples and the iPod oranges—that need to be addressed. To make alternate energy a reality by adapting existing technology or developing new technology will require the kind of technological well-funded push afforded reluctantly to Samuel Morse for his electromagnetic telegraph, advocated by Vannevar Bush in his "Endless Frontier" essay, or promised NASA by President Kennedy. It just may have received at least some of that funding by way of The American Recovery and Reinvestment Act of 2009 that went into effect in February of 2009. Within the hefty $790 billion economic stimulus package are provisions to pump tens of billions of dollars into battery research and manufacturing by way of grants, loans, and tax incentives for everything from manufacturing plants to basic research. Whether this is enough to play catch-up with well-funded competitors in Asia remains to be seen, but it is a start.

ON A MUCH SMALLER SCALE, MIT researchers are experimenting with microbatteries about half the size of a human cell. However, it isn't the size of the battery that has generated interest, it is the assembly process. A genetically altered virus called M13 is set loose on a specially prepared surface to build up material for the anode. This is the kind of research that could lead to incredibly small self-powered ICs for implantable sensors.

And then there are so-called bio-batteries. As early as 2003, students at St. Louis University in Missouri developed a battery that ran on alcohol, specifically vodka and gin, using a catalyst to break down the components into enzymes. Sony followed suit with a battery that runs on sugar. Like the vodka-gin battery, enzymes break down or digest the sugar (glucose) while a specially designed anode extracts the electrons and hydrogen ions from the sugar. Not long ago in Singapore, a scientist unveiled a small bio-battery that uses urine as an electrolyte. Although it garnered a fair share of jokes in the media, the so-called pee battery essentially runs on the same principle as the power source in the proximity fuse—just add electrolyte—and could find applications in powering low-energy medical tests—for example, a diabetes or pregnancy test.

At Rensselaer Polytechnic Institute, scientists have come up with a somewhat similar concept, a battery that uses bodily fluids as an electrolyte. Paper-thin, the battery is comprised primarily of cellulose with nanotubes printed on it. The advantage, the researchers pointed out, is that it's easily cut to size to fit comfortably under the skin for a device such as a pacemaker.

BATTERY DEVELOPMENT IS, AT LONG last, catching up to related fields. Today, the basic chemical principles that generate power for a vast range of devices are more alike than they are different. Tomorrow, they will likely become as different as the devices they power. Science and technology adapt, change, and interact in often surprising ways. More than two centuries ago, a dead frog's leg unexpectedly twitched. The debate that followed has long been settled, but the science that emerged continues today and will no doubt continue far into the future.

# Appendix

## Those Troublesome Baghdad Batteries

The mystery began with a flood. In 1936 snowmelt running off the mountains in Turkey flowed downward into small tributaries and streams before entering the Tigris River; snaking its way south through Iraq, the river overflowed its banks before entering the low-lying floodplain and joining the Euphrates on its journey to the Persian Gulf. The floodwaters divided the Iraqi capital, turning the eastern section of Baghdad into an island.

The large stagnant pools that remained in the low-lying areas alarmed public health officials. Fearing an outbreak of malaria, a plan was devised to remove a small loamy hill from an area near Khanaqin in eastern Iraq and fill in the stubborn ponds of standing water. However, almost as soon as work crews began to excavate the mound, they discovered the remains of ancient dwellings. The Baghdad Archaeological Museum was alerted and a small expedition hastily assembled to collect the artifacts. It might have ended there with a modest, though welcome, find, had it not been for Wilhelm Koenig (or König).

An artist by training, Koenig arrived in Iraq from Germany in 1930 with an archaeological expedition and by 1936 found himself a curator at the Baghdad Archaeological Museum. If Koenig's career move from artistic landscapes to archaeology seems odd, it suited the relatively new museum in a country whose borders were also newly drawn.

Among the artifacts uncovered at the Khujut Rabuah dig was a small broken clay jar with a cylinder made from a rolled sheet of copper and an iron rod. The small, bright yellow oblong jar, about 5 inches high and 3 inches in diameter, was a puzzling thing for Koenig. As with the other artifacts removed from the mound, including several bowls used in rituals, Koenig dated it from the Parthian era, around 200 BC to 200 AD.

What possible use could it have held for the ancient people? Similar jars found in the early 1930s in nearby Seleucia were said to be containers for sacred scrolls. Koenig, however, took another approach. Through reverse engineering, he theorized that the mysterious piece of pottery might have been a battery and wrote up his findings after returning to Germany in 1939.

> The answer to the question as to the use of the curious find caught me by surprise when I brought all of the parts into relationship with each other and considered their careful separation from each other by insulating asphalt: It must be an electric battery! One need only to put in an acid or alkaline liquid and the battery would be finished. I expressed my view with caution, as it could only be confirmed by further circumstances of discovery and discoveries . . .

Koenig addressed the most obvious question—the purpose of the battery—with an electroplating theory. Other artifacts discovered in the region seemed to be electroplated. These included bronze and copper vessels covered with a flaky patina. And, too, local craftsmen of the Parthian era used a somewhat unique method of applying a

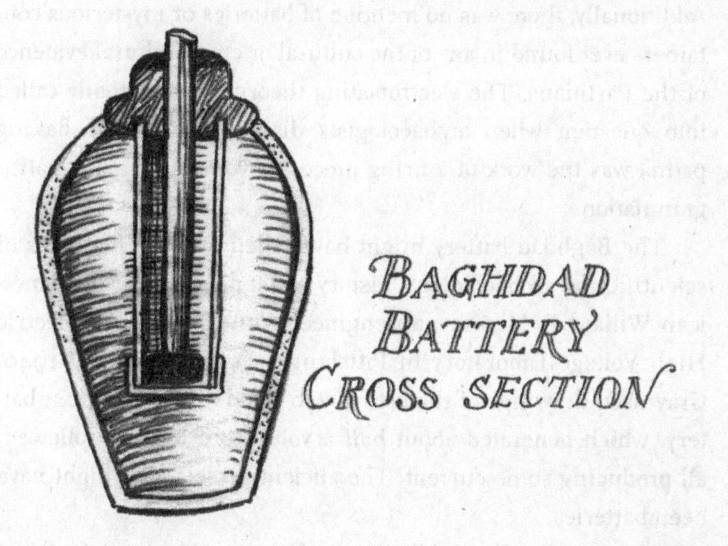


metallic veneer in a rudimentary form of electroplating, which may have evolved from a more ancient form. Electricity generated from the jar might also have been used in religious ceremonies. It may have also had a medicinal purpose, since the ancient Greeks had used the electric charge from torpedo fish as a local anesthesia.

Over time more of the mysterious jars, some with a slightly different design, were found, about a dozen in all, which Koenig theorized might have been linked together to generate a more powerful current. Not surprisingly, Koenig's battery theory was widely rejected by the archaeological establishment. There were also problems with the theory from a technological standpoint. The top of the container was sealed with bitumen, an asphaltlike substance, a design flaw that left whoever used the battery unable to replace the acidic or alkaline electrolyte. Would an ancient people have designed a disposable battery?

Other evidence mounted against the battery theory. Although the Parthians maintained a substantial empire for a time, they did not appear to have been particularly sophisticated technologically.

Additionally, there was no mention of batteries or mysterious containers ever found in any of the cultural or cross-cultural evidence of the Parthians. The electroplating theory was eventually called into question when archaeologists discovered that the flaking patina was the work of a firing process involving mercury, called granulation.

The Baghdad battery might have fallen into the footnotes of scientific and archaeological history had it not been for the American Willard F. M. Gray, an engineer with the General Electric High Voltage Laboratory in Pittsburgh, Pennsylvania. In 1940, Gray used drawings of the artifacts to build his own ancient battery, which generated about half a volt. More replicas followed, all producing some current. The ancient vessels just might have been batteries.

For even the most coldly skeptical, the combination of ancient ingenuity and modern mystery is intellectually enticing, except for all those nagging technical details. For instance, to perform any work, such as electroplating, several of these low voltage batteries would have to be linked in series—positive to positive/negative to negative—to boost the voltage. Yet no such linking apparatus was found. However, perhaps what is most worrisome about the Baghdad battery theory is that it lacks any form of known underlying science in the Parthian culture.

The Baghdad batteries stand alone as a sophisticated artifact in an otherwise perfectly consistent ancient culture. If the Parthians had created a battery, time has obscured its purpose as well as every trace of its scientific origins. Part of the problem, of course, is the very simplicity of battery technology itself—two dissimilar metals in an acidic or alkaline solution. A copper penny and a galvanized nail pushed through the skin of a lemon creates an electrical charge. The Parthians might well have stumbled on the secret of generating electricity by accident.

The mystery of the Baghdad batteries has endured for more

than seventy years. A broken piece of yellowish pottery about the size of a man's fist continues to surface as proof of the wisdom of the ancients and to support the myths of pop culture junk science. UFO enthusiasts have even claimed it as evidence of interplanetary visitors, while others have cited it as evidence of interdimensional travelers.

## *Author's Note*

Careful readers will no doubt find a number of battery applications that have either not been included or are only touched on lightly. The reason is simple: space and time. Batteries are so ubiquitous in our modern world that it would be impossible to include all of the devices they power. There is also the issue of word usage when it comes to "energy" and "power." For purposes of style, they are used nearly interchangeably with apologies to the more technically minded reader. Finally, the author is well aware that the Metallica song, "Battery," from the 1986 *Master of Puppets* album used as a chapter epigraph is in all likelihood not referencing power sources, but rather, the concept of "assault and battery" as well as San Francisco's Battery Street. However, the temptation to include it between the same covers as excerpts from Emily Dickinson, Lord Byron, Herman Melville, et al., was simply too good to pass up.

# Selected Bibliography

Ackerman, Kenneth D. *The Gold Ring: Jim Fisk, Jay Gould, and Black Friday, 1869.* New York: Harper Business, 1988.

Allen, Frederick Lewis. *Only Yesterday: An Informal History of the 1920's.* New York: Harper & Row, 1931; Perennial Classics, 2000.

"The Atlantic Cable, Successful Completion of the Great Work." *New York Times,* July 30, 1866.

Benjamin, Park. *History of Electricity (The Intellectual Rise in Electricity) From Antiquity to the Days of Benjamin Franklin.* New York: John Wiley & Sons, 1898.

"Bio-battery Runs on Shots of Vodka." *New Scientist,* March 24, 2003.

Bodanis, David. *Electric Universe: How Electricity Switched on the Modern World.* New York: Three Rivers Press, 2005.

Bode, Carl, and Malcolm Cowley, eds. *The Portable Emerson.* Rev. ed. New York: Viking, 1981.

Bragg, Melvyn, with Ruth Gardiner. *On Giants' Shoulders: Great Scientists and Their Discoveries from Archimedes to DNA.* New York: John Wiley & Sons, Inc., 1998.

Brinkman, William F., Douglas E. Haggan, and William W. Troutman. "A History of the Invention of the Transistor and Where It Will Lead Us." *IEEE Journal Of Solid-State Circuits* 32, no. 12 (December 1997).

Brodd, Ralph J. *Factors Affecting U.S. Production Decisions: Why Are There No Volume Lithium-Ion Battery Manufacturers in the United States?* Prepared for Economic Assessment Office Advanced Technology Program National Institute of Standards and Technology, June 2005.

Bullock, Kathryn R. "Samuel Ruben: Inventor, Scholar, and Benefactor." The Electrochemical Society *Interface*, Fall 2006.

Burke, James. *The Pinball Effect: How Renaissance Water Gardens Made the Carburetor Possible—And Other Journeys through Knowledge*. Boston: Little, Brown and Company, 1996.

Byron, George Gordon, Baron. *Byron: Poetical Works*. Edited by Frederick Page. 3d ed. Corrected by John Jump. Oxford and New York: Oxford University Press, 1970.

*Cambridge Dictionary of Science and Technology*. Peter M. B. Walker, general editor. Cambridge: Cambridge University Press, 1988.

Castells, Manuel, and Mireia Fernández-Ardèvol, Jack Linchuan Qiu, and Araba Sey. *Mobile Communication and Society: A Global Perspective*. Cambridge, Mass.: MIT Press, 2007.

Cobb, Cathy, and Harold Goldwhite. *Creations of Fire: Chemistry's Lively History from Alchemy to the Atomic Age*. New York: Plenum Press, 1995.

Collins, A. Frederick. *Wireless Telegraphy: Its History, Theory, and Practice*. New York: McGraw Publishing, 1905.

*Concise Dictionary of World History*. New York: Macmillan Publishing Company, 1983.

Coulson, Thomas. *Joseph Henry: His Life and Work*. Princeton: Princeton University Press, 1950.

Cutcliffe, Stephen H., and Terry S. Reynolds, eds. *Technology & American History: A Historical Anthology for Technology & Culture*. Chicago: University of Chicago Press, 1997.

Davis, L. J. *Fleet Fire: Thomas Edison and the Pioneers of the Electric Revolution*. New York: Arcade Publishing, 2003.

Donat, James G. "The Rev. John Wesley's Extractions from Dr. Tissot: A Methodist Imprimatur." In *Brain, Mind, and Medicine: Essays in Eighteenth Century Neuroscience*. Edited by Henry Whittaker, C. U. M. Smith, and Stanley Finger.

Dubpernell, George. "Evidence of the Use of Primitive Batteries in Antiquity." In *Selected Topics in the History of Electrochemistry*, edited by George Dubpernell and J. H. Westbrook. Princeton: The Electrochemical Society, 1978.

"E. A. Calahan, Inventor, Dies, Originator of Gold and Stock Ticker and Multiplex Telegraph System." *New York Times*, September 13, 1912.

Edgerton, David. *The Shock of the Old: Technology and Global History since 1900*. New York: Oxford University Press, 2007.

Edwards, Marvin R. "Joan and Eleanor: Radio Transmissions aboard the Mossie." *OSS Society Newsletter* 8 (Fall 2006).

"Electric Girls." *New York Times,* April 26, 1884.

"The Electrical Display, Nothing Like It Has Ever Been Seen in the World—The Search Lights." *New York Times,* May 13, 1893.

"Electrical Show Open: The Exhibition at the Garden Inauguration by President McKinley from Washington." *New York Times,* May 3, 1898.

*Electro Importing Company: Catalogue No. 7,* First Edition 1910

*Encyclopædia Britannica: A Dictionary of Arts, Science and General Information,* 11th ed. New York: The Encyclopædia Britannica Company, 1910.

Fahie, J. J. *A History of Wireless Telegraphy.* New York: Dodd, Mead & Company, 1901.

"Fairchild Introduces Circuits in Miniature." *New York Times,* March 15, 1961.

Fenichell, Stephen. *Plastic: The Making of a Synthetic Century.* New York: HarperCollins, HarperBuisness, 1996.

Findlen, Paula, ed. *Athanasius Kircher: The Last Man Who Knew Everything.* New York & London: Routledge, 2004.

Fitzgerald, F. Scott. *The Great Gatsby.* New York: C. Scribner's Sons, 1925; Scribner Paperback Fiction, 1995.

Fleming, A. J. *The Principles of Electric Wave Telegraphy and Telephony.* New York: Longmans, Green, and Company, 1916.

Fowler, Gene, and Bill Crawford. *Border Radio: Quacks, Yodelers, Psychics, and Other Amazing Broadcasters of the American Airwaves.* Austin: University of Texas Press, 2002.

Friedel, Robert D. *Lines and Waves: Faraday, Maxwell and 150 Years of Electromagnetism.* New York: Institute of Electrical and Electronics Engineers, 1981.

Gilbert, William. *De Magnete.* Translated by P. Fleury Mottelay. New York: Dover Publications Inc., 1958.

Grant, John. "Experiments and Results in Wireless Telephony." *The American Telephone Journal,* January 26, 1907.

Gribbin, John. *The Fellowship: Gilbert, Bacon, Harvey, Wren, Newton, and the Story of a Scientific Revolution.* Woodstock, N.Y.: Overlook Press, 2007.

Hayes, Jeremiah, F. "Paths Beneath the Seas: Transatlantic Telephone Cable Systems," *IEEE Canadian Review,* Spring/Printemps, 2006.

Hawthorne, Nathaniel. *The House of the Seven Gables.* Edited by Robert S. Levine. 2d ed. New York: W. W. Norton, 2005.

Heilbron, J. L. *Electricity in the 17th and 18th Centuries: A Study in Early Modern Physics.* Berkeley: University of California Press, 1979; Mineola, N.Y.: Dover Publications Inc., 1999.

Hospitalier, E. *The Modern Applications of Electricity.* Translated and enlarged by Julius Maier. New York: D. Appleton & Co., 1882.

"An Incomplete Display; The Electrical Exhibition Still in Confusion." *New York Times,* September 4, 1884.

Israel, Paul. *Edison: A Life of Invention.* New York: John Wiley & Sons, Inc., 1998.

James, Frank A. J. L., ed. *"The Common Purposes of Life": Science and Society at the Royal Institute of Great Britain.* Aldershot, U.K.: Ashgate Publishing Ltd., 2002.

———. *The Development of the Laboratory: Essays on the Place of Experiment in Industrial Civilization.* New York: American Institute of Physics, 1989.

James, Peter, and Nick Thorpe. *Ancient Inventions.* New York: Ballantine Books, 1994.

Jones, Richard Foster. *Ancients and Moderns: A Study of the Rise of the Scientific Movement in Seventeenth-Century England.* 2d ed. St. Louis: Washington University Press, 1961.

Kahn, David. *The Codebreakers: The Story of Secret Writing.* New York: Macmillan, 1967.

Kawamoto, Hirohisa. "The History of Liquid-Crystal Displays." *Proceedings of the IEEE* 90, no. 4 (April 2002).

Knight, David. *Humphry Davy: Science & Power.* Oxford: Blackwell, 1992; Cambridge: Cambridge University Press, 1996.

Krause, Reinhardt. "Inventor of Portable Radio Developed Joan-Eleanor Project." *OSS Society Newsletter,* Fall 2006.

Linden, David, and Thomas B. Reddy, Ph.D., "Battery Power for the Future: Is the Energy Output of Batteries Reaching Its Limit?" *Battery Power Products & Technology,* March/April 2008.

———. *Handbook of Batteries.* 3d ed. New York: McGraw-Hill, 2002.

"Marvelous Electrical Inventions Displayed, Attractions at a 'Conversazione' at Columbia University." *New York Times,* April 13, 1901.

Marvin, Carolyn. *When Old Technologies Were New: Thinking about Electricity Communication in the Late Nineteenth Century.* New York: Oxford University Press, 1988.

Massie, Keith, and Stephen D. Perry. "Hugo Gernsback and Radio Magazines: An Influential Intersection in Broadcast History." *Journal of Radio Studies* 9, no. 2 (2002).

McElroy, Gil. "A Short History of the Handheld Transceiver." American Radio Relay League, *QST*, January 2005.

Melville, Herman. *Moby-Dick*. Edited by Hershel Parker and Harrison Hayford. 2d ed. 150th anniversary ed. New York: W. W. Norton, 2002.

"MIT Engineers Work toward Cell-sized Batteries." *MIT News*, August 20, 2008.

Moore, Gordon E. "Cramming More Components onto Integrated Circuits." *Electronics* 38, no. 8 (April 19, 1965).

Morse, Samuel. *Modern Telegraphy, Some Errors of Dates of Events and of Statement in the History of the Telegraphy Exposed and Rectified*. Paris: A. Chaix and Co., 1867.

Munro, J. *Heroes of the Telegraph*. London: Religious Tract Society, 1891.

Murphie, Andrew, and John Potts. *Culture and Technology*. New York: Palgrave Macmillan, 2003.

"A New Way to Fire Mines; A Successful Exhibition of Electrical Transmission without Wires in the Garden." *New York Times*, May 7, 1898.

"The News of Radio, Two Shows on CBS Will Replace 'Radio Theatre' During the Summer." *New York Times*, July 1, 1948.

Nye, David E. *Electrifying America: Social Meanings of a New Technology, 1880–1940*. Cambridge, Mass.: MIT Press, 1990.

O'Malley, Michael. *Keeping Watch: A History of American Time*. New York: Viking, 1990.

O'Neill, Eugene. *Long Day's Journey into Night*. New Haven: Yale University Press, 1989.

Odean, Kathleen. *High Steppers, Fallen Angels, and Lollipops: Wall Street Slang*. New York: Holt, 1988.

Orrine, Dunla, Jr. "Under the 'Big Top'; New Radio Outfits to Be Displayed Sept. 20–30 at New York Electrical Show." *New York Times*, September 3, 1933.

Pancaldi, Giuliano. *Volta: Science and Culture in the Age of Enlightenment*. Princeton: Princeton University Press, 2003.

Partington, J. R. *A Short History of Chemistry*. 3d ed. New York: Dover, 1989.

Peña, Carolyn Thomas de la. *The Body Electric: How Strange Machines Built the Modern American*. New York and London: New York University Press, 2003.

Pera, Marcello. *The Ambiguous Frog: The Galvani-Volta Controversy on Animal Electricity*. Translated by Jonathan Mandelbaum. Princeton: Princeton University Press, 1992.

Petrus Peregrinus. *The Letter of Petrus Peregrinus on the Magnet, A.D. 1269*.

Translated by Brother Arnold. New York: McGraw Publishing Company, 1904.

Pope, Frank L. *Modern Practice of the Electrical Telegraph: A Handbook for Electricians and Operators.* 7th ed. New York: Van Nostrand, 1872.

Press Release from Bell Telephone Laboratories, July 1, 1948.

Pushparaj, Victor L., Manikoth M. Shaijumon, Ashavani Kumar, Saravanababu Murugesan, Lijie Ci, Robert Vajtai, Robert J. Linhardt, Omkaram Nalamasu, and Pulickel M. Ajayan. "Flexible Energy Storage Devices Based on Nanocomposite Paper." *Proceedings of the National Academy of Sciences of the United States of America* 104, no. 34 (August 21, 2007).

"The Regency Transistor Radio, TR-1." *Consumers' Research Bulletin,* July 1955.

Rejan, Wendy. "Fort Monmouth Engineers Get Smart!" *Monmouth Message,* September 28, 2007.

Rhodes, Richard. *The Making of the Atomic Bomb.* New York: Simon & Schuster, 1986.

Richardson, Alan. *British Romanticism and the Science of the Mind.* Cambridge: Cambridge University Press, 2001.

Riordan, Michael. "How Europe Missed the Transistor." *IEEE Spectrum,* November 2005.

Riordan, Michael, and Lillian Hoddeson. *Crystal Fire: The Birth of the Information Age.* New York: W. W. Norton, 1997.

Ruben, Samuel. *Necessity's Children: Memoirs of an Independent Inventor.* Portland, Ore.: Breitenbush Books, 1990.

Sack, Edgar A. "Consumer Electronics: An Important Driver of Integrated Circuit Technology." *Proceedings of the IEEE* 82, no. 4 (April 1994).

Sacks, Oliver. *Uncle Tungsten: Memories of a Chemical Boyhood.* New York: Alfred A. Knopf, 2001.

Schallenberg, Richard H. *Bottled Energy: Electrical Engineering and the Evolution of Chemical Energy Storage.* Memoirs series 148. Philadelphia: American Philosophical Society, 1982.

Schindall, Joel. "What's in a Name? A New Model for Regenerative Electrical Energy Storage." *Power Electronics Society Newsletter* 20, no. 1 (First Quarter 2008).

Scuilla, Charles J. "The Commercialization of Lithium Battery Technology," The Battcon International Stationary Battery Conference, 2007.

Shelley, Mary. *Frankenstein: or, The Modern Prometheus: The 1818 Text.* Edited by

Marilyn Butler. New York: Oxford University Press, 2002.

Shima, Masatoshi. Interview by William Aspray. May 17, 1994. Interview # 197, The Institute of Electrical and Electronics Engineers, Inc. IEEE History Center, and Rutgers, The State University of New Jersey.

Singer, Charles, E. J. Holmyard, A. R. Hall, and Trevor I. Williams, eds. *A History of Technology*. Vol. 4, *The Industrial Revolution c. 1750–1850*. New York: Oxford University Press, 1958.

Standage, Tom. *The Victorian Internet: The Remarkable Story of the Telegraph and the Nineteenth Century's On-line Pioneers*. New York: Walker, 1998.

Stevenson, William R. "Miniaturization and Microminiaturization of Army Communications—Electronics, 1946–1964." Headquarters U.S. Army Electronics Command, Fort Monmouth, New Jersey, Historical Monograph 1, Project Number AMC 21M.

"The Storage of Electricity, the Subject Explained by Prof. George F. Barker." *New York Times*, March 31, 1883.

Tassey, Gregory. "Standardization in Technology-Based Markets." National Institute of Standards and Technology, June 1999. Forthcoming in *Research Policy*.

Taylor, Robert Lewis. "Profiles: High Railers and Full Scalers." *The New Yorker*, December 13, 1947.

Taylor, William B. *A Memoir of Joseph Henry, A Sketch of His Scientific Work*. Philadelphia: Collins, 1879.

"Terminology for Semiconductor Triodes." Memoranda. Bell Telephone Laboratories, Inc. May 28, 1948.

Thomas, Dana L. *The Plungers and the Peacocks*. New York: Putnam, 1967.

Thoreau, Henry David. *Walden*. 150th anniversary illustrated edition. Boston: Houghton Mifflin, 2004.

Truffaut, François, with Helen G. Scott. *Hitchcock*. New York: Simon & Schuster, 1967. Originally published as *Le cinéma selon Hitchcock*, 1966.

Uglow, Jenny. *The Lunar Men: Five Friends Whose Curiosity Changed the World*. New York: Farrar, Straus, & Giroux, 2002.

Untermeyer, Louis. *Makers of the Modern World: The Lives of Ninety-two Writers, Artists, Scientists, Statesmen, Inventors, Philosophers, Composers, and Other Creators Who Formed the Pattern of Our Century*. New York: Simon & Schuster, 1955.

Van Doren, Carl. *Benjamin Franklin*. New York: Viking, 1938.

"War Work Seen at Electric Show." *New York Times,* October 11, 1917.

Weightman, Gavin. *Signor Marconi's Magic Box: The Most Remarkable Invention of the 19th Century & the Amateur Inventor Whose Genius Sparked a Revolution.* Cambridge, Mass.: Da Capo Press, 2003.

Whittemore, Katharine, ed. *The World War Two Era: Perspectives on All Fronts from Harper's Magazine.* New York: Franklin Square Press, 1994.

Wightman, William P. D. *The Growth of Scientific Ideas.* New Haven: Yale University Press, 1953.

Wireless Telegraph Advertisement: Telimco. *Scientific American,* November 25, 1905, page 427.

Wittenberg, Gunter. "Obituary: Paul Eisler." *The Independent,* October 29, 1992.

Worts, George. F. "Directing the War by Wireless." *Popular Mechanics,* May 1915.

Zheludev, Nikolay. "The Life and Times of the LED—A 100-Year History." *Nature Photonics* 1 (April 2007).

# Index

Page numbers in *italics* refer to illustrations.